VOLUME ONE · WEST

COTTIAN ALPS

01 / Bivacco Valmaggia
02 / Bivacco Bonelli
03 / Bivacco Danilo Sartore
04 / Bivacco Barenghi
05 / Bivacco di Punta Venezia
06 / Bivacco Tornior
07 / Bivacco Lago Laus
08 / Bivacco Robinet
09 / Bivacco Berrone
10 / Bivacco Vallonetto
11 / Bivacco del Sommeiller
12 / Bivacco Mario Sigot
13 / Bivacco Walter Blais
14 / Bivacco Hannibal
15 / Refuge du Camp des Rochilles

DAUPHINÉ ALPS

16 / Refuge de Chalance
17 / Habert des Sabottes
18 / Habert de la Pierre
19 / Habert des Jarlons

SAVOY PREALPS

20 / Chalets Forestiers de la Balme
21 / Chalet du Charbon
22 / Chalet de Cotagne

GRAIAN ALPS

23 / Rifugio Santa Maria
24 / Bivacco Beltrando
25 / Bivouac d'Estelette – Adolfo Hess

VOLUME TWO · EAST

GRAIAN ALPS

26 / Bivacco Gervasutti
27 / Bivacco Cesare Fiorio
28 / Bivouac du Dolent

PENNINE ALPS

29 / Bivacco Brédy
30 / Bivacco Pansa
31 / Bivacco Rivolta
32 / Bivacco Tzan
33 / Mischabeljochbiwak
34 / Cresta-Biwak

BERNESE ALPS

35 / Fusshornbiwak
36 / Aarbiwak

LEPONTINE ALPS

37 / Bivacco Zeb
38 / Bivacco Petazzi / Lago Ledù

RHAETIAN ALPS

39 / Bivacco Anghileri-Rusconi
40 / Bivacco Giampaolo Del Piero
41 / Bivacco Città di Cantù

DOLOMITES

42 / Bivacco Donato Zeni
43 / Baita di Cece
44 / Bivacco Forcella Coldosè
45 / Bivacco Paolo e Nicola
46 / Bivacco Carlo Minazio
47 / Bivacco Sperti
48 / Bivacco Ugo Dalla Bernardina
49 / Bivacco Bocco-Zago
50 / Bivacco de la Medassa

VOLUME ONE · WEST

CABIN WILD

Alpine Bivouac Huts

Mediterranean to Mont Blanc

Ben Tibbetts & Valentine Fabre

PREFACE

Scattered across the European Alps are more than 600 tiny, open-access cabins. This constellation of unmanned huts offers simple shelter in some of the wildest corners of these mountains.

The Alps are a complex and layered landscape, where towering glaciated peaks lie in close proximity to densely populated, economically developed areas. As a result, generations of herders, soldiers, smugglers and mountaineers have left their mark on the range. Over time, this steep and complex terrain has been threaded with trails and dotted with an eclectic patchwork of manmade structures: shepherds' cabins, barracks, border posts and climbing shelters, each shaped by particular needs and distinct historical contexts.

While it is tempting to view the Alps primarily as a recreational space, these modern values are somewhat at odds with the deeper history of the sylvo-pastoral economy, a way of life rooted in livestock herding and forestry. For millennia, alpine life was defined by subsistence rather than by leisure. Over the last two centuries, shifting values, technologies and economic pressures have drastically reshaped both the physical landscape and its cultural significance.

The persistent tension between traditional land use and evolving environmental priorities continues to surface in new forms. The controversy surrounding the return of the grey wolf to the western Alps epitomises this delicate balance. Having been eradicated from the Alps in the 1940s, this apex predator, protected as part of biodiversity conservation efforts in recent decades, has come into conflict with livestock farmers. To guard their herds, many now rely on aggressive Pyrenean Mountain Dogs, which in turn pose risks to hikers. It's a poignant contradiction: efforts to preserve one strand of natural heritage risk alienating those defending another. The mountains nurture competing truths.

Front cover: David Lee by Bivacco Beltrando at sunrise, Val Soana, Italy.
Back cover: Bivacco Hannibal at dusk, Valle di Susa, Italy.
Right: Bivacco Berrone at dusk, Valle di Susa, Italy.

Our book focuses on unguarded refuges, celebrating their diversity, simplicity and beauty, both as cultural artefacts and as springboards for deeper exploration. Many lie in a liminal zone between the working, pastoral landscape and the mineral world of the high mountains. With so many competing agendas and ideals, it is something of a miracle that many of these structures exist at all. Although the perspective we present is largely recreational, my background – growing up on a sheep farm, then training as a historian, and later as a mountain guide – hopefully offers a view from both sides of the fence.

For nearly a decade, 4000-metre peaks shaped the adventures Valentine and I pursued. While we were writing and translating *Alpenglow*, we noticed that the journeys that lingered longest in our memories weren't the most famous or accessible lines, but the routes further from the beaten track, often accessed from a humble bivouac hut.

Over time, our focus shifted to the lesser-known corners of the Alps, where small huts became both destination and departure point. Together, we've crossed the range on foot, by bike and on skis, not in search of comfort or convenience, but seeking the wildest places and the most compelling lines. These shelters gradually became the framework that linked our adventures.

Our project grew from that impulse: to explore the variety of these architectural forms, to photograph them and piece together their origins. We invite you to explore the traces left by generations before us. The paths they carved traverse not only physical space but layered histories: of agrarian labour, military presence and mountaineering desire. In this context, each hut is not just a shelter, but a marker of human presence and imagination in the high Alps.

Their presence on alpine climbing routes has long stirred debate. As early as the 1930s, Lucien Devies, president of the French Alpine Club, denounced bivouac shelters as "réducteurs d'aventure" – diminishers of adventure. Though safety has often trumped such purist ideals, the question remains: what kind of mountain experience do we wish to protect?

This question lies at the heart of our approach. We recognise the paradox in publishing a book about quiet places. Yet through careful curation, practical advice, and an emphasis on local culture and historical context, we aim to encourage deeper, more respectful exploration rather than add to the noise. With that intention, these first two volumes present 50 shelters, spanning the Alps from southern France to the Italian Dolomites. They were selected not for their comfort or prestige but for the breadth of experience they represent.

Behind each of these structures lies a story and a community. Built and maintained largely by volunteers and local groups, these huts bring joy and, at times, safety to countless visitors each year. This fragile, generous network endures thanks to goodwill, and deserves respect and care. These shelters carry cultural significance as beacons of hospitality, resilience and alpine heritage. They offer a lens through which to view the Alps as a shared space shaped by history and imagination. Whether you're seeking primitive stone huts or experimental architecture, embarking on simple hikes or technical mountaineering routes, these cabins open up a spectrum of possibilities.

Left: Valentine near Habert de la Pierre at sunset, Belledonne, France.

CONTENTS

Right: Valentine riding over Col des Cerces, Massif des Cerces, France.

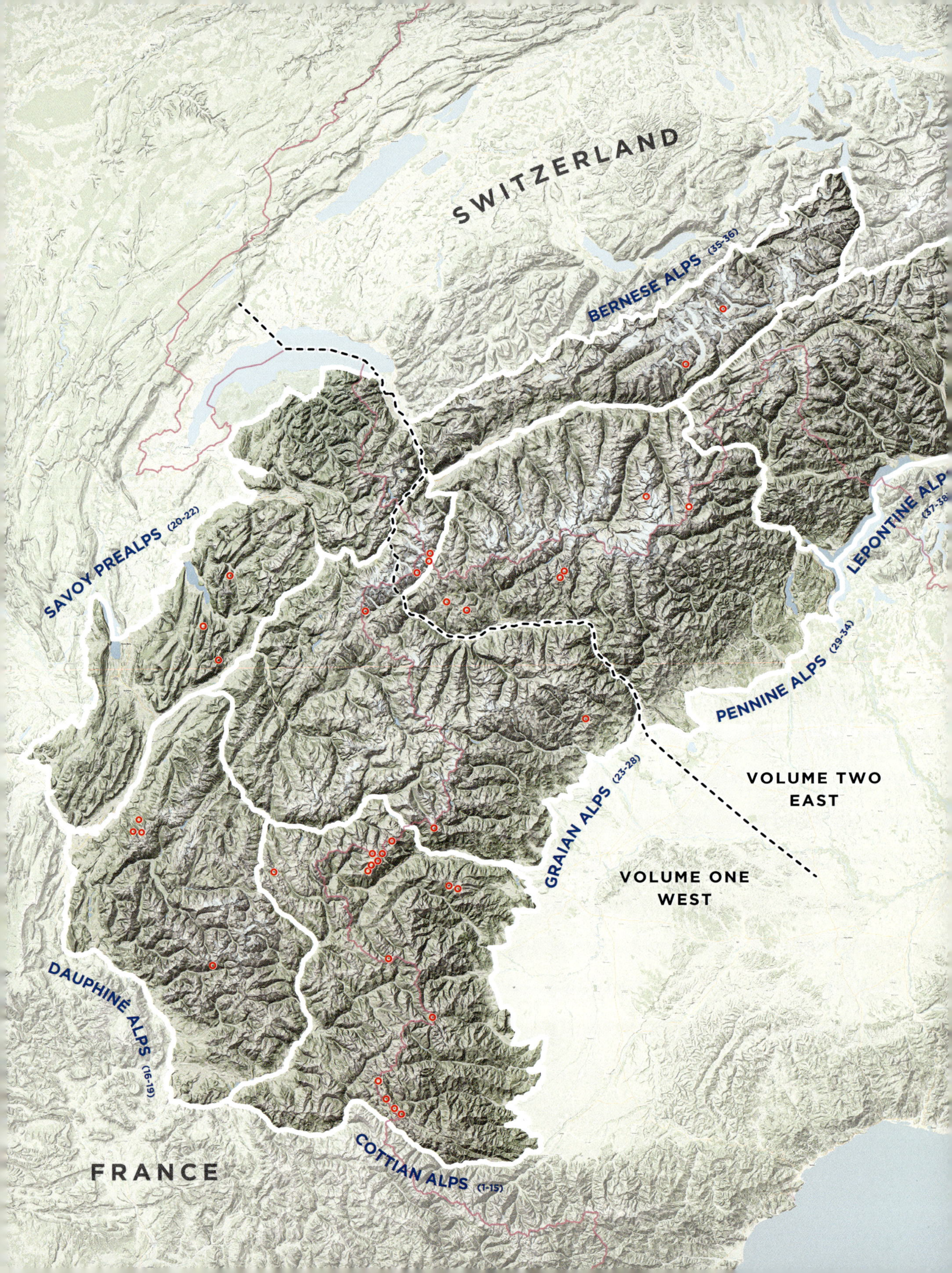
SWITZERLAND
BERNESE ALPS (35-36)
SAVOY PREALPS (20-22)
LEPONTINE ALPS (37-38)
PENNINE ALPS (29-34)
VOLUME TWO
EAST
GRAIAN ALPS (23-28)
VOLUME ONE
WEST
DAUPHINÉ ALPS (16-19)
FRANCE
COTTIAN ALPS (1-15)

DOLOMITES (42-50)
RHAETIAN ALPS (39-41)
ITALY

Introduction – Shelter, Story and Connection

Scotland, Antarctica and the Alps

Laden with coal and kindling, I set out with a small group of friends into the desolate heart of the Cairngorms on New Year's Eve. Drawing over the bare, bouldery summit of Càrn a' Chlamain, shrouded in dense fog, we pored over the map, hunting for the easiest line down, curving northwards into this remote corner of the Scottish Highlands. Warm, wet winds had stripped the mountains, leaving only isolated patches of snow – stranded coral islands in a sea of heather.

We bounded down the grassy gully of Fèith Uaine Bheag. The mist drained the landscape of colour and form but enveloped us in musty, acidic scents of peat bog and ling heather. As the gully flattened into the broad basin, we dipped below the cloud. Anticipation rose in my throat, the excitement and intrigue of not knowing what, or whom, we might encounter. It had been years since I last visited this corner of wilderness. Amid the patchwork of heather and grass tussocks, a prehistoric palette of umber and ochre, I thought such vivid moments would never fade. For years afterwards, I could recall every bend in the stream and roll of hillside, but now, twenty years later, time has frayed the edges. The few photos I took have become crutches for my memory.

In such a barren, storm-lashed landscape, a damp stone hut feels like a haven of peace. Six hours from the road, we were nearing Fèith Uaine Bothy. Set on a marshy mountain plateau at the confluence of four streams, it is one of the most remote huts in Britain. The current building, a former hunting lodge, was rebuilt from the ruins of dwellings dating back to the 17th century. The light was fading into the long northern night as we rose over a bluff, stumbling over peat hags and spongy moss as the hut came into view, its long green roof hung over low, mossy walls merging into the brown, frost-bitten grass. I lingered on the far side of the stream, watching as Richard and Lettie clipped their boots against the step, eased open the door and disappeared inside.

Above and left: Richard McDonald, Hal Jacob, Tom Owen and Andy Barlow in Fèith Uaine Bothy (Tarf Hotel), Cairngorms, Scotland.

I had been living in a third-floor apartment above a busy street in Edinburgh. Day and night, as the buses below slowed to a stop, the building tremored like the city's slow heartbeat, pulsing every twenty minutes. Having grown up on a farm before moving north to study at art school, city life had been a startling change. Escaping to the hills soon became a refuge from the cacophony of modern life. Getting out into open country felt like a necessity.

As the last light drained from the sky, I tried to etch the scene into my memory: the stream, recently in spate, had left a mosaic of ice chunks in a neat line along the grassy banks, beach wrack marking the high tide. The clouds drifted close overhead, briefly parting to reveal the sky. It seemed unlikely that anyone else would arrive, so we would have the hut to ourselves, our ship adrift in a sea of moorland. Crossing the threshold, the damp stone walls held the bitter aroma of decades of woodsmoke. As my friends brewed tea, I shrugged off my colossal pack, stuffed a pawful of malt loaf in my mouth and set about lighting the fire. The isolation of a bothy has a way of intensifying relationships, and as I lay in my sleeping bag that night, listening to rain hammering against the tin roof, I felt content, surrounded by friends and far from the clamour of the city.

—

I learnt haphazardly how to climb, first on the small sea cliffs and quarries near Edinburgh, then in the rugged mountains of the Highlands, from the Cairngorms and Ben Nevis to the Cuillin on the Isle of Skye. Roaming these vast, weather-beaten landscapes, I soon learnt of the network of mountain huts, or "bothies", Scotland's counterpart to the alpine bivouac hut. Back then, the whereabouts of these revered structures were scarcely published, revealed only in a cryptic list of names and coordinates available to members of the Mountain Bothy Association (MBA).

The MBA's mission statement captured my imagination: "To maintain simple shelters in remote country for the use of all who love wild and lonely places." I pored over that list, gathering second-hand maps covering the entire country and tracing out routes: hikes, climbs and biking loops. These gradually formed a network of journeys that, over ten years, would lead me to nearly every bothy in Scotland. Just as arbitrary as trying to climb all the 3000-foot peaks, seeking out these remote shelters became my way of discovering the country and pursuing a quiet personal goal.

Scattered across the ancient Caledonian hills, these humble stone structures offer basic shelter to weary adventurers. Many were abandoned crofts, remnants of a subsistence culture forced off the land during the Highland Clearances of the 18th century. Later they were renovated to house shepherds or ghillies on vast estates. These isolated buildings now stand sentinel on barren hillsides and deserted coastlines, accessible only after hours of strenuous hiking. The irony is not forgotten that the brutality of the Clearances eventually led to a network of secluded, romantic refuges, as wilderness emerged from what were once thriving communities.

Bothies became my introduction to the historical tapestry of the Scottish wilderness, a lens through which I came to understand the landscape, and a destination for adventures shared with friends. There was a heady romance to the idea: walking for miles across boggy ground, past remnant shielings and ruined townships abandoned two centuries earlier, battered by wind and rain, only to arrive at a tiny hut. Yet beneath the lichen and moss, the barbaric history was barely hidden. That sense of wilderness, empty and untouched, was but a mirage.

Nevertheless, reaching a bothy always felt like coming home. It revived that childhood instinct to build shelters, dens and hideouts in the woods. Against the noise of urban life, these remote refuges stirred something primal in me: a longing to disappear into the mountains, to live like a hunter, shepherd or smuggler, a fugitive from an overbearing civilisation. This attraction to incongruous shelters in wild landscapes, like an oasis in the desert, seems hardwired into our very nature, perhaps an echo of ancestral survival instincts. As small, naked apes ill-suited to the extremes of mountain environments, we are drawn to places of refuge.

—

Leaving Scotland, I moved to Switzerland for two years. I spent more time in the high mountains, climbing classic routes and visiting Alpine bivouac huts with exotic names: the Bouquetins, Rossi e Volante and l'Envers des Dorées. Drawn by the lure of the great white wilderness, I headed south in 2009 to work in Antarctica for 18 months. During the austral summer, as I gradually adapted to the frozen landscape, I was assigned to work at several remote huts, research stations built during successive eras of polar exploration. Although the surrounding landscape felt atemporal, these isolated outposts were steeped with the stories of those who had dwelled there before me. What started for me as names and coordinates on the vast map of the Antarctic Peninsula, soon evolved into narrative and history. Each shelter was a chapter in the ongoing story of human interaction with this wilderness.

While Antarctica's history is less violent than Scotland's, these stations had been established, abandoned and repurposed as research priorities shifted, or as climate conditions altered access. I spent a month at Fossil Bluff, a wooden hut at 71° south, overlooking George VI Sound, a vast frozen fjord up to 1000 metres deep. Built in 1961, it had once housed three men through eight long winters in the 1960s and 70s, stranded in the darkness with only HF radios for communication. Even in 2009, a year's supply of dried food still sat in storage, reminding us that we too might find ourselves stuck if the Twin Otter planes broke down. Working over the frozen sea sometimes gave me vertigo, my mind unable to ignore the depth of ice beneath my feet. As the summer progressed, meltwater pooled on the surface. Shimmering under the low sun, it gave the illusion that the fjord had already thawed, as if human-triggered climate change had already run its course.

Stationed at Rothera that winter, a modern research station on the edge of Adelaide Island, the sun barely rose for months. As I adjusted to the long polar nights, my fascination with bivouacs was rekindled. I dreamt of the freedom to roam across mountain ranges, unburdened by polar survival equipment.

Top: Fossil Bluff Field Station, Alexander Island, Antarctica. Bottom: Sky-Blu Field Station, Palmer Land, Antarctica.

Across a painfully slow internet connection, I began piecing together a route to traverse the European Alps, seeking out the most remote and beautiful corners of that vast yet accessible mountain range.

In almost every region, I discovered small, unguarded refuges: bivouac huts of all shapes and sizes. Although I had spent a decade enamoured with the austere charms of Scottish bothies, the idea of exploring Alpine bivouacs felt like stepping onto a grander, more dramatic stage, where glaciers and towering rock walls replaced the mist-shrouded glens of home.

—

Unlike modern guarded refuges, which are gradually transforming into comfortable high-altitude hotels, offering duvets, showers and hot meals, most unguarded bivouac shelters in the Alps remain true to their original purpose: basic places of shelter and staging posts for the next day's adventure. From the late 19th century, many huts began employing wardens not only to provide food and comfort for guests, but also to protect the buildings from vandalism. However, the more remote and inaccessible cabins saw too few visitors to warrant even a seasonal custodian. Today, bivouacs that are easy to reach still suffer from the carelessness and degradations of an inebriated minority. Despite this, many Alpine clubs and associations, particularly in Italy, continue to place remarkable faith in their fellow humans, maintaining a vast network of shelters that remain open and accessible.

Like Scotland's bothies, Alpine bivouacs are remarkable symbols of cultural generosity. The men and women who invest the time, resources and imagination into creating and maintaining these structures rarely stay in them; rather, they hope that others will find refuge or joy. Such a phenomenon stands in stark contrast to the commodification of modern tourism, where experiences are packaged, priced and consumed. An unguarded bothy or bivouac, by virtue of its openness, remains a space of trust in which every visitor is both guest and caretaker.

Beyond their immediate utility, bivouac huts invite reflection on broader themes of architecture, geography and how humans inscribe themselves on the environment. In a world where most things serve a clear function, building a tiny shelter at 3000m can seem irrational, or even idealistic. Yet it is precisely this sense of the improbable that fuels the allure. As such, bivouacs are akin to the architectural follies of past centuries, expressions of passion rather than necessity.

With few pre-existing models to emulate other than the shanties and lean-tos of the shepherds and crystal hunters, Alpine bivouacs have become spaces of radical experimentation. Scholars and architects have long recognised that high-altitude construction pushes the boundaries of building practices. Design competitions for new huts – judging proposals on architectural quality, sustainability, energy efficiency and technical feasibility – have proved fertile ground for exploring how we might inhabit the natural environment with a reduced footprint. For engineers and architects alike, each new hut becomes a laboratory of structural innovation, where materials like fibreglass, metal and specialised composites are tested against the ravages of altitude. Yet despite these advancements, wood remains the primary structural material in most recent projects, valued for its versatility, durability and low environmental impact.

The construction of these shelters is itself an adventure in architecture. Parallel to the history of alpinism, the history of shelter-building in the high mountains has been equally audacious, marked by masterful success stories and catastrophic failures. The environment is often extreme: in the highest locations temperatures regularly drop below -20°C and winds exceed 100km/h. With abundant rain and snowfall, site selection is critical, balancing the risks of rockfall, flooding and avalanches. Moreover, as global temperatures rise and permafrost degrades, vast swathes of terrain are now becoming at risk from landslides.

In the 19th and 20th centuries, human intrusions into European mountains were seen as symbols of progress and modernity. The construction of mountain railways, cable-cars, vast ski resorts and hydroelectric schemes was widely celebrated. However, this trend has been sharply reversed in recent years as we begin to appreciate the fragility and intrinsic value of the Alpine landscape.

Left: Valentine in the Rifugio Quintino Sella, Mont Blanc, Italy. The floor tilts, so rocks under the table legs keep it level.

The concept of reversibility, building new huts in ways that allow for their removal with minimal environmental scarring, reflects shifting cultural values that prioritise ecological stewardship over permanent human occupation. Today, refuges are among the few human structures still welcomed in these environments. As attitudes shift, the huts we build, the paths we tread and the stories we tell all contribute to the larger narrative of our relationship with the land. In the words of J.B. Jackson, "landscape is history made visible."

In these lofty realms, the line between architecture and art installation has begun to blur. Some recent huts stand as bold sculptural statements that emphasise counterpoint: a gleaming modern volume pitched amid rugged, primeval peaks. Most, however, pursue a more discreet integration into the landscape, seeking to blend with rock strata or echo crystalline contours. The tension between robust functionality and near-poetic minimalism lies at the heart of what makes mountain bivouacs so compelling. Unlike the mass-produced tin huts of the mid-20th century, which were placed in all manner of strategic locations, most contemporary cabin design enters into an explicit and figurative dialogue with the territory and the natural environment.

As we move further into the 21st century, alpine climbing routes and their attendant shelters face growing uncertainty. Climate change is triggering extreme weather events and accelerating the retreat of permafrost. Faced with these concerning developments, each hut stands as a testament to human resilience: our enduring drive to find safety, express creativity and foster community in a spectacular but unforgiving arena.

In an age where adventure is increasingly quantified, distilled into speed records and numerical achievements, the experience of a night in the mountains defies such reduction. Bivouac huts offer something intangible. More than mere logistical assets, they embody a philosophy of engagement with the land, one rooted in humility, adventure and a shared respect for the environment we inhabit. *Cabin Wild* is a celebration of these remarkable places. It is an invitation to explore not only the huts themselves, but also the histories, landscapes and human stories woven into their existence.

In August 2024, a decade after climbing past it in the night, Valentine and I finally made the long-overdue pilgrimage to Rifugio Quintino Sella, a stone-clad wooden cabin perched at 3371m on a spur protruding from the west face of Mont Blanc. This unguarded shelter, one of the oldest still standing, was built in 1885 to facilitate the ascent of the Tournette Spur. At that time, before the discovery of the Pope route, the Tournette was considered the easiest way to the summit of Mont Blanc from Italy.

We had heard rumours that the rock beneath the hut had shifted due to permafrost melt and that the structure was in jeopardy. After a six-hour approach, we arrived just before sunset to find part of the stone wall bulging outwards. Yet, anchored by thick steel cables, the hut held fast, and we risked heading inside. The floor slanted at a dizzying fifteen degrees and the wooden structure was warped. Rectangular sections had become rhomboids. The entire hut was twisted, its eastern half leaning precariously into the void.

Despite the obvious risk of being there, I was fascinated by this historic monument, perhaps in its final days. Compelled to explore every nook and cranny, I sought a connection to the legendary climbers who had passed before me. The wooden interior felt like an archive of late 19th-century exploration. Before the light faded, I photographed its details: the sturdy wooden beams once carried up by hand and the inscriptions of renowned guides etched on the walls, some dating back to the hut's opening year.

As we stood on the threshold at dusk, jagged peaks looming overhead, we became acutely aware of our fragility and of how dependent we are on the simplest forms of shelter, and thus on one another. Perhaps that is the ultimate reason these huts spark our imaginations: they offer not only refuge, but stories and human connections.

Above and left: Rifugio Quintino Sella on the west face of Mont Blanc, Italy.

Practicalities

Overview

Bivouac huts are typically basic, with few or no facilities. Most lack stoves, heating, electricity or running water, offering only a weatherproof shelter and a place to sleep. Many provide mattresses and woollen blankets, though this isn't guaranteed. Although many unguarded huts remain open year-round, some, particularly in Switzerland, have an online reservation system and may even require a key to be collected from a local shop or town hall. Check the information section in each chapter and, where possible, verify current details with the local alpine club, guides' bureau or tourist office before your trip.

Clothing

In addition to your hiking, skiing or climbing gear, consider packing an extra down jacket and insulated trousers in case the hut is colder than expected. A spare base layer and socks are always welcome. Even if blankets are provided, bring a sleeping bag liner to help keep bedding fresh and hygienic.

Water

Water sources in the Alps are often safe to drink if they are drawn from a spring or a fast-flowing stream with no nearby grazing animals. However, avoid using water downstream of a bivouac, refuge or any other habitation. If in doubt, boil or filter your water. Some bivouac huts have no nearby water source – while melting snow can be an option in winter and spring, in summer you may need to carry water from the starting point. Wash your dishes with biodegradable detergent and pour the wastewater on the ground, ensuring it doesn't flow directly back into the watercourse. The nearest water supply is noted in each hut's description.

Food

Dehydrated meals and dried foods like pasta, couscous and noodles are lightweight and quick to cook. Tea bags and coffee are ideal for hot drinks. Many Italian bivouacs have a moka pot, although it often needs a good wash. For breakfast, muesli or cereal bars work well, while sandwiches, bars and chocolate are handy for lunch. To keep mice away, hang food supplies from the ceiling or place them under an inverted pan.

Cooking

Most bivouacs lack cooking facilities, and even those with stoves may be unreliable. It's wise to carry a small gas stove, lightweight pan and lighter as a backup. In winter, an efficient stove with a large pan is essential to melt enough snow to stay hydrated.

Wood & Stoves

We have noted whether each bivouac has a woodstove or fireplace. Some high-altitude huts, especially in Switzerland, receive wood supplies by helicopter, but most do not. Please use supplied wood sparingly, as it's costly to transport.

Likewise, avoid making outdoor fires, as they waste resources, damage the ground and encourage others to follow suit. Be prepared to carry fuel from the valley, or gather dry deadwood in the last woodland area on your approach. Huts near the treeline may have an axe and saw, but not always, so carrying a folding saw can be helpful. Avoid cutting live branches, as this harms trees and green wood burns poorly. Whenever possible, collect extra deadwood to leave for others in case they arrive in distress. Before lighting a woodstove, ensure safe and efficient operation: check that the chimney isn't blocked, the flue is airtight and the door seals properly. When emptying the ash drawer, ensure there are no live embers, pieces of metal or plastic waste, then spread the ashes at least 20m from the hut. Lastly, it is worth noting that some recently constructed huts are so well sealed and insulated that it's essential to cook outside or ensure adequate ventilation to prevent carbon monoxide poisoning from an inefficiently burning gas stove.

Toilets

Very few bivouacs have a toilet. When nature calls, choose a spot at least 300m from the hut, downhill from the site, well away from paths and at least 500m from a water source to prevent contamination. In summer, use a trowel or the heel of your boot to dig a hole, or lift a few stones to bury your waste in rocky terrain. Whenever possible, use leaves, moss or snow instead of toilet paper. If toilet paper is essential, burn it (unless there's a wildfire risk), or bag it and carry it out.

In winter, it's even more important to go far from the hut and ensure you're not near a summer path. Ideally, dig down to the ground; if that's not possible choose a discreet site. Remember that anything left on the snow will eventually resurface in spring. Small handfuls of compressed snow make an excellent alternative to toilet paper in winter. Excrement around popular bivouacs is a significant problem, often due to a lack of foresight or imagination. Please don't contribute to it!

Tent & Sleeping Bag

I usually carry a tent or bivvy bag as a backup when visiting a bivouac, especially in summer. Even if all bed spaces are taken, floor space is often available. In emergencies, latecomers should always be accommodated. However, in some small huts, such as the Eccles on the south face of Mont Blanc, there is no extra room, and on fine summer days, demand far exceeds capacity. In such cases, bringing the right gear to sleep outside is essential. Overcrowded huts can be uncomfortable even if you can find a spot indoors, so always have a solid backup plan, whether sleeping outside or retreating to the valley. In winter, it's wise to carry a sleeping bag; even if blankets are provided, they can be damp and not very warm.

Right: David Lee preparing to leave Bivacco Beltrando, Punta Rama, Piemonte, Italy. Previous spread: Bivacco Hannibal at night, Valle di Susa, Italy.

Essential Items (Depending on Hut Equipment)

Rucksack, clothes, hat & gloves, waterproofs, sleeping bag, silk/cotton liner, sleeping mat, food, trash bag, stove, gas, lighters, pan, utensils, cup, bowl, map, compass, whistle, sunglasses, phone (with mapping apps), headlamp, knife, water bottle, water filter, toiletries, sunscreen, first-aid kit, earplugs, toilet paper, alcohol gel. Backup items: bivvy bag or lightweight tent.

Transport

While reaching the start of these itineraries is often simpler by car, all the valleys are accessible by public transport with a bit of planning, and this liberates you to travel end to end, rather than looping back to your starting point. Switzerland's transport network is excellent, but France and Italy's are less comprehensive, so you may want to hitchhike or take a taxi from the last town to the trailhead.

Maps & Navigation

Many of these huts are located in high-mountain terrain, where navigation can become challenging in poor weather. Before venturing into the mountains, ensure you're proficient in navigation and carry a detailed map of the area along with a compass. While phone GPS mapping software is now highly accurate, it's crucial to have a backup and know how to use it. For maps, IGN is recommended for France, Swisstopo for Switzerland and nearby Italian areas, Fraternali for Italy's Piemonte and Aosta regions, and Tabacco for eastern Lombardy, Trentino-Alto Adige and Veneto.

Weather & Mountain Risks

Before heading into the mountains, study the weather forecast carefully and prepare for sudden changes. Even with a clear outlook, mountain weather can shift unexpectedly, so always have backup plans and the humility to turn back if necessary.

Etiquette

Bivouac huts exist thanks to the generosity of their owners; please respect this privilege. While formal rules vary by country, the core principles remain: be considerate to others, care for the hut, respect the environment and keep group sizes small.

Most of the high-altitude shelters in the Alps were built by alpine clubs to support climbers. Priority at these huts is given first to those in danger or distress, then to mountaineers preparing for a route the next day, and lastly to hikers. In some regions, such as Austria, bivouac huts are intended only for emergency use, rather than casual stays. In all unguarded huts, visits should be limited to one or, at most, two nights. If you're seeking a quieter experience, aim for an area with a long approach and, if possible, avoid weekends and the peak summer months of July and August.

Consideration for Others: Treat all visitors with respect. Welcome others, share space and ensure your group doesn't disrupt other people's experiences. Smoking is prohibited in huts. Always make room for anyone arriving in perilous conditions. For smaller bivouacs, avoid your team filling more than half the beds. Large groups (over six) will likely detract from the experience of other users, so should seek the usage of a private hut or guarded refuge instead. Commercial trekking groups should seek special arrangements with owners. Generally, the shelters described are open to all, but may be locked without warning for maintenance, private use or when assigned to a shepherd.

Respect for the Environment: The most common problem in bivouacs is visitors leaving behind rubbish and human waste. Carry out all your rubbish, including wine bottles – bivouacs don't need more "candle-holders". Whenever possible, remove litter left by others. Respect local regulations: many huts are located in national parks, nature reserves or other protected areas where wild camping is forbidden, and visitors must follow marked itineraries in both summer and winter. Avoid lighting fires, picking wildflowers or making excessive noise that could disturb wildlife or fellow hikers. In many Alpine regions, dogs must be kept on a leash, and in some areas, they are forbidden altogether – check the rules before you go.

Respect for the Hut: Before leaving, fold blankets, sweep up, and leave the hut as you'd wish to find it. Prevent fire hazards – extinguish flames, turn off gas and electricity, and securely close doors, windows and shutters before leaving. Never leave graffiti or place stickers on a hut, and if you find any damage or mess, do your best to address it, then report any outstanding problems to the hut owner, local alpine club, guides' bureau or tourist office. Leave cash in the provided box (where supplied) or make a bank transfer to support the hut's upkeep.

Backup Plans & Unforeseen Situations: Always have an alternative plan or carry the necessary gear to sleep outside in case a hut is closed, full or unsuitable. In rare cases, you may encounter inconsiderate groups treating a bivouac as a party spot. If that happens, your best option may be to camp elsewhere or descend.

Sanitation: Never leave human waste anywhere near huts. If you encounter such issues, help preserve these spaces for future users and attempt to clean up and dispose of waste far from the site. This effort can prevent owners from deciding to lock up the hut for good.

Rapid ascents above 3000m may trigger acute mountain sickness, with symptoms ranging from headaches and nausea to life-threatening oedema. On snow, reflected sunlight can cause severe sunburn and snow blindness. In winter, the risks multiply, with avalanches and frostbite becoming greater threats. Caution, preparation and proper training are essential for navigating these environments safely.

Avalanche

While generally sheltered from avalanches, most of these huts have approaches that cross avalanche terrain when snow is present. Avalanches are perhaps the most complicated hazards to avoid and the most difficult risk to evaluate, so restraint and prudence are required until settled snow conditions prevail.

Emergencies

When heading into the mountains, always leave a route description and expected return time with a friend. If someone in your party is injured, follow first-aid procedures to stabilise the situation. Determine your exact location and call for help if you have reception – dial 112 in EU countries or 144 in Switzerland.

Mountain guides typically carry emergency VHF radios or satellite communication devices, which could aid in getting assistance. Make sure everyone in your group is aware of the plan and is able to communicate with each other if separated. Assess your options, and if necessary, have someone walk out to get help if there's no mobile signal.

Left: Valentine studying the map in Bivacco Mario Sigot, Massif du Mont-Cenis, Italy.

Information, topo and map key

04	*Bivouac number*
🚌 P	*Public transport destination or parking for approach*
🛏	*Bed places (number of mattresses and blankets)*
🔥	*Cooking or heating facilities*
🚰	*Water sources*
i	*Additional information*
↗ ⌂	*Approach time and distance to the hut*
↗ ▲	*Mountain climb from the hut*
↻ ▲	*Mountain tour passing by the hut*
⛷	*Ski tour from the hut*

———	Line of approach or route
– – –	Hidden part of the route
— · —	Variation or alternative route
III+	Technical grade of a climbing section
○	Refuge, bivouac, lift station or belay
○↓	Rappel
(B)→	Feature mentioned in the text

Itinerary Information

Although this book is primarily a collection of photographs, stories and destination ideas, additional itineraries are suggested for each hut, many of which we've personally undertaken. Below are indications on how to interpret and use this information. Each bivouac chapter begins with a grey information box detailing the road access point, number of beds and blankets, cooking facilities and water supply. The normal access route is described, followed by a selection of nearby adventures. Many of these routes require significant experience to be undertaken safely. Moreover, both hut infrastructure and the mountain terrain are constantly evolving, so some details may become outdated over time.

Itinerary Length & Timing

The stated length for each itinerary indicates the cumulative distance or vertical gain required. Suggested timings are based on a fit, acclimatised party with appropriate experience for the route, in good conditions.

Hiking Grades

For hiking routes, difficulty levels are indicated using the Swiss Alpine Club hiking scale:

T1: Hiking – The path is well marked. The terrain is regular and any exposed sections are very well secured. Navigation remains simple and all risk of accident can be avoided with normal behaviour.
T2: Mountain hiking – Parts of the path can lack markings. The terrain can be steep at times, with some risk of falling. Basic navigation skills and good footwork are required.
T3: Demanding mountain hiking – The trail may not always be visible in the terrain and exposed sections may be secured with ropes or chains. Hands might be needed for balance. Some stretches might be exposed to rockfall or cross moderately steep rocky slopes. Mountain experience, good footwork and navigation skills are necessary.
T4: Alpine hiking and scrambling – At times there is no path and hands may be needed to progress. The terrain can be exposed, with delicate grassy slopes, rocky inclines, easy snowfields and sections of bare glacial ice (no crevasse danger). Participants must have experience of exposed terrain, good terrain assessment skills and strong navigation abilities. Alpine experience is necessary, as retreating in bad weather can be challenging.
T5: Demanding alpine hiking and scrambling – This terrain often lacks a defined path and includes sections of easy climbing. It can be extremely exposed, with rocky slopes, glaciers and snowfields presenting a risk of slipping. Considerable alpine experience is essential, along with basic proficiency in using an ice axe, crampons and rope. Confident terrain assessment and excellent navigation skills are also required. T5 can overlap with the climbing grade "F". While the Swiss hiking scale extends to T6, this level increasingly overlaps with the climbing grades "F" and even "PD", as described below.

Alpine Climbing Grades

For most technical routes, especially where guidebook information is scarce, detailed descriptions are provided. For some harder routes, where comprehensive details are readily available, only a summary is provided.

For alpine climbing, we have used the French adjectival system to give a guide to the overall challenge that will be encountered on each route when undertaken in favourable or normal conditions. This grade is subjective but considers the required technical skill, the sustained nature of difficulties, as well as the route's length, complexity, protection, commitment and objective hazards:

F: facile (easy) – Straightforward glacier trips or snow climbs up to 40°, where ropes are necessary; simple rock scrambles with minimal exposure or technical challenges.
PD: peu difficile (slightly difficult) – Glacier expeditions with crevasses or steeper sections; short, exposed snow slopes up to 45–50°; narrow ridge lines without significant technical demands; extended rock scrambles and shorter rock routes with occasional pitches of grade II/II+.
AD: assez difficile (moderately difficult) – Snow and ice climbs with slopes up to 55° often requiring belays; sustained narrow snow ridges needing good balance and crampon skills; intricate rock ridges needing proficient scrambling techniques in ascent, descent and traverse; rock routes with pitches up to III/III+ and occasional IV moves. Rappels may be necessary.
D: difficile (difficult) – Steep snow and ice climbs up to 60° with possible mixed climbing on icy rock; long, exposed and delicate snow ridges; major rock or mixed ridge climbs with sustained sections of IV/IV+ and short pitches of V or V+.
TD: très difficile (very difficult) – Extensive snow and ice routes with complex navigation or technical challenges and pitches up to 85°; rock or mixed ridges and faces with sustained difficulties of V and sections of VI.
ED: extrêmement difficile (extremely difficult) – Major face or couloir climbs with vertical ice pitches and/or long, serious mixed sections, often with significant objective hazards; sustained rock face and buttress routes with many pitches of VI and above.

On climbs with rock passages, we have noted the relevant UIAA technical rock grades (Roman numerals) or the French rock grades (Arabic numerals).

Ski Grades

For ski terrain, the grade is provided using the "Toponeige" system. The rating includes five levels: 1.x to 5.x. The first four levels have three subdivisions (x.1, x.2 and x.3) while the 5th level is open-ended. The ratings are as follows: Ski 1.x: wide slopes <30°; Ski 2.x: slopes <35°; Ski 3.x: presence of technical passages with long slopes at 35° and short sections up to 40–45°; Ski 4.x: couloir or steep skiing with slopes at 40 or 45° for more than 200m; Ski 5.x: slopes at 45°/50° for more than 300m or more than 50° for 100m.

Left: sunset on the south ridge of Pointe de Banc Plat, Massif des Bauges, France. Following spread: Finlay Wild and Valentine Fabre by Bivacco Danilo Sartore, Val Maira, Italy.

COTTIAN ALPS

19/ Habert des Jarlons
18 / Habert de la Pierre
17 / Habert des Sabottes
Modane
Valloire
15 / Refuge des Rochilles
Bardonecchia
Le Bourg-d'Oisans
La Grave
D A U P H I N E
A L P S
Briançon
16 / Refuge de Chalance
La Chapelle-en-Valgaudémar
L'Argentière-la-Bessée
Guillestre
Gap
Embrun

23 / Rifugio Santa Maria
11 / Bivacco Colle del Sommeiller
14 / Bivacco Hannibal
13 / Bivacco Walter Blais
Susa
12 / Bivacco Mario Sigot
10 / Bivacco Vallonetto
09 / Bivacco Berrone
Torino
07 / Bivacco Lago Laus
08 / Bivacco Robinet
Sauze d'Oulx
Sestriere
06 / Bivacco Tornior
TTIAN
ALPS
05 / Bivacco di Punta Venezia
04 / Bivacco Barenghi
03 / Bivacco Sartore
02 / Bivacco Bonelli
01 / Bivacco Valmaggia
Cuneo

The Skyline of Valle Maira

Bivacco Valmaggia · Bivacco Bonelli · Bivacco Danilo Sartore · Bivacco Barenghi

Across the southern Alps, even in mid-winter, the Mediterranean air acts like a blowtorch, stripping snow from the south-facing slopes between each storm cycle. This provides a feast of fresh grass for the ibex and chamois, whose numbers are barely checked by the growing wolf population. The wolves, it seems, prefer fresh lamb, plucked from the abundant flocks just across the French border. While the south-facing slopes lie bare, the north-facing valleys can still hold snow well into spring, remaining accessible only on skis or snowshoes.

For years, Valle Maira had remained just an idea, a name whispered among mountain connoisseurs. Friends spoke of it as one of the mythical corners of the Alps, where the mountains are quiet and the sun always shines. Visiting during a February anticyclone only reinforced this notion. Like much of Piemonte, this remote valley has no lift infrastructure. With long approaches to little-known peaks, this rough, complex landscape naturally filters out all but the most passionate bipeds.

With few guarded refuges in this region, several chilly nights in bivouac huts were on the menu. Valentine and I gathered provisions and set out with two of our fittest friends, Finlay Wild and Ben Bardsley. Both were UK-based fell runners. Ben and I had teamed up for many ski-mountaineering races, with him often dragging me to the finishing line with an elastic bungee. His quadriceps were legendary across the racing community, where he is known as 'The Quadfather'. Fin, quiet and reserved, has a wiry frame and an infectious grin. After months at home in the wet Scottish Highlands, he arrives in the Alps each season buzzing with energy and ideas.

We set off from Campo Base, at the far end of Valle Maira, weighed down with heavy packs. At Colle dell'Infernetto, we caught our first glimpse

Left: Valentine Fabre, Ben Bardsley and Finlay Wild above Bivacco Sartore. Above: the SE ridge of Auto Vallonasso at sunrise from Bivacco Valmaggia.

into the region's heart, Monte Sautron in the distance and the chiselled top of Rocca Blancia. After crossing two more cols, we raced up the valley below Brec de Chambeyron, chasing the last light on Bivacco Barenghi.

Given its remote location and the time of year, we were surprised to find the small blue cabin already occupied. A cold breeze picked up and we threw on more clothes while we assessed the situation. Shams, a guide from Grenoble, his partner and their three clients had already settled in. They peered out over the stable door at us, hardly managing to hide their dismay. Sham's first words, before even greeting us, were: "You know there's a refuge not far from here on the French side?" It was a weird moment.

There were enough beds for all of us, but the tiny hut promised a cramped evening. We shoe-horned ourselves onto the remaining benches and made a brew. Sipping tea together, we exchanged stories and discovered shared passions, itineraries and acquaintances. Despite the benevolence of Sham's group, playing cards while quietly defending their prime spots around the table, it felt claustrophobic. As dusk approached, we made the snap decision to break camp and ski across the French border to the Refuge du Chambeyron. I have rarely felt so liberated to leave a hut and head out into the night.

With no clear information about the refuge, we took a gamble. Nevertheless, after an hour of thrashing through crusty, refrozen snow, we crested a ridge and spotted lights glowing in the hut. We arrived to find the stove blazing, and the refuge warmed by three friendly skiers. Much as I love spartan bivouac huts, in winter temperatures, a heated refuge was a welcome upgrade from the frigid metal can of Bivacco Barenghi.

I realise I sometimes take for granted our liberty to roam in the Alps and the freedom to access a welcoming network of refuges and bivouacs. A few months later, our experience in North America highlighted just how lucky we are in Europe. In late spring, having climbed Denali, the highest mountain in North America, Valentine and I had grown weary of freezing in a tent at high altitude, so we set out along the Resurrection Pass Trail in the rolling Kenai Mountains of southern Alaska.

Late one evening, we arrived at a beautifully built, two-storey log cabin beside the idyllic Trout Lake. The only sound was the eerie howl of loons as they skimmed the trees and landed on the water. Despite being a "Public Use Cabin", owned by the National Forest Service with space for eight, we were turned away by two aggressive elderly women. In that region, a single person can reserve the exclusive use of a cabin six months in advance, and most of the season gets booked up the moment reservations open. Every hut we passed was already reserved online, yet the only other person we met was another canvas-bound vagrant like ourselves.

The downside of the European "first come, first served" bivouac ethic is the risk of arriving late to a full hut. I usually carry a tent as a backup, but in Valle Maira, in mid-winter with a fit team and plenty of other cabins nearby, our backup plan was to keep moving until we found space. Two bivouacs had caught my attention: Bivacco Bonelli, set spectacularly above a lake against the jagged backdrop of Tête de Moïse (Monte Oronaye in Italian), and the newer Bivacco Danilo Sartore, notable for its stylish design.

After a long, comfortable night in Refuge du Chambeyron, we set out slowly under a milky sky, crossing back over the frontier past Bivacco Barenghi. As we skinned south across undulating terrain, the wind harried us, driving snow flurries along the ground. By late morning, we reached Col de la Portiola, just west of Monte Sautron. A large, old male ibex surveyed us from the ridge as we snuck back into France, escaping the biting northerly breeze.

Guarding the east face of La Meyna, golden limestone turrets caught the emerging sun. But our eyes were drawn to something less natural, the stark geometric form of the Batterie de Viraysse. This fortification, a high-altitude outpost of the colossal Forte de Tournoux, was built in the mid-19th century to defend the Ubaye valley from Italian invasion.

Most Alpine fortifications were well concealed, and a century later, they have begun to blend into the mountainside. However, the angular lines of the Viraysse Fort remain well defined in the winter landscape, despite damage sustained during the Second World War. I could almost picture the bustle of soldiers forging a life up here year-round. Standing at 2772m, Viraysse housed an artillery battery and barracks for 450 men. In the 19th century, it was the highest French fortification, constructed in a strategic position overlooking the Vallée de l'Ubayette where the road descends from the frontier at Col de Larche (Colle Della Maddalena in Italian).

For millennia, this was a strategic trade and military passage, one of the easier trans-Alpine crossings. Today, it is among the few road passes kept open through winter. The first road was built in 1515 by the army of the French king François I, paving the way for 40,000 troops to invade Italy.

When reading about the history of a mountain region, I'm intrigued by what I can't see in the landscape, what has been concealed by man or erased by time. Yet the military vestiges I encounter invoke a quiet horror, the unease of someone who has only known war as a historical abstraction. There is a dissonance between the wild-looking terrain and the innumerable concrete bunkers dissimulated into the mountainside; their presence stirs a morbid fascination. I wonder if future generations will view our industrial infrastructure – reservoirs, tunnels and pylons – with a similar unease, as material remnants of our quest to conquer nature.

During the Second World War, the high valley before us was the stage for a brief but intense combat in the First Battle of the Alps. By May 1940, with German forces already overwhelming northern France, Mussolini seized the moment. On 10 June 1940, Italy declared war on France and Britain. As the French ambassador François-Poncet observed, it was "a dagger blow to a man who has already fallen."

Right: Valentine and Ben ascending the northeast couloir of Tête de Moïse.

On 17 June, the first of Mussolini's forces crossed Col de Larche. Three days later Italian heavy artillery bombarded the Viraysse Fort and the following day Italian infantry broke through the high mountain passes at Col de Sautron and Col des Monges to approach Viraysse. The battle raged for several days. Although Viraysse was never overrun, the village of Larche was captured by the Italians. On 24 June 1940, France signed an armistice with Italy in Rome, ending the fighting. In the following weeks, the forts in this region were disarmed, and their artillery was sent north to reinforce other fronts.

As the high cloud dispersed, leaving only wisps of cirrus, the harsh light revealed a rugged and deserted landscape buried under snow. The fort and other relics stood like surreal exclamations in the mineral topography, a reminder of how fortunate we are to live in a time and place of relative peace. The day warmed rapidly as we skied up Col de Sautron. Wind and sun had scoured the snow, exposing barbed wire tangled in the scree and grass. Beyond the pass, a ribbon of scrappy snow led us down to the incongruous form of Bivacco Danilo Sartore, a red triangular cabin standing beside ruins from the Second World War. A mouse would need to be hardy to survive on the rations left behind, let alone a famished walker, faced with several kilos of salt and a packet of coffee.

We pressed on over another pass into the stunning hanging valley of Lago d'Apsoi, dominated by the steep shattered north face of Tête de Moïse. Skating across the undulating ground, our packs swinging with each stride, we finally glimpsed Bivacco Bonelli just as the shadow line crept towards it. Racing down the last slope, we reached the hut five minutes before the sun slipped behind the ridge. In the fading light, I worked intensely on a small ink drawing, but before I could finish, the mountains dissolved into shadow.

The following morning, we skinned up through a frigid reservoir of cold air to the base of Tête de Moïse.

Above: Finlay climbing the west ridge of Tête de Moïse. Top right: Finlay by Bivacco Valmaggia at sunrise. Bottom right: Valentine skiing below Col de Feuillas.

The mountain was slow to reveal itself, but a steep, diagonal couloir cleaved the north face, leading to the western shoulder. We climbed in crampons as the couloir narrowed to an awkward chute just four metres wide, hemmed in between crumbling rock walls. Above us, the gully opened up, leading more easily to the sunny col. While Valentine and Ben (Bardsley) descended the warm snow of the south couloir, Fin and I clattered in carbon ski boots along the western ridge, over a few short buttresses and along the crest to the ornate summit cross.

Rejoining the others at Col de Feuillas, we side-slipped our way down the steep, icy gully below the pass. Looking up to our left, a long, flat concrete building lay almost entirely buried in snow, Casermetta del Feuillas, a former barracks tucked just below the frontier. Once a base for Italian troops stationed on this strategic peak, it has since been renovated into an austere shelter, Bivacco Enrico e Mario. Above, a via ferrata completed in the 1930s ascends the craggy southeast face of Cima Dronero, threading past two lookout posts that lie camouflaged among the towers and gullies. This route, the only via ferrata in the Western Alps built for the Second World War, formed part of the "Vallo Alpino", Italy's Alpine Wall defensive system, designed as a counterpart to France's Maginot Line. Maintaining these remote military outposts, especially during winter, proved to be a perilous undertaking.

Leaving the weight of history behind, we continued down on cold, feathery snow, making our way to Bivacco Valmaggia, the wooden hut in the valley below. The hut was spacious, warm and comfortable, everything we could wish for in the depths of winter.

After tea and a late lunch, Fin was still itching for more activity, and his infectious enthusiasm caught on. With darkness approaching, we headed up the steep slopes towards Bric Content, a sharp peak rising above the hut. Hypnotised like moths, we were drawn to the last sunlight on the ridge. A strip of snow led upwards to the loose, rocky crest and a surprisingly exposed little summit. We didn't linger. As the light faded, we skittered back to our skis and made careful turns down the steep terrain to the hut.

On our last morning, I rose early and drank my tea outside, watching the sunrise roll down the mountainside towards us. A warm spring breeze drifted across the valley, caressing the snow. It was going to be a hot day, and it was time to leave. As we set off, a troop of chamois gazed down from a steep spire, knocking off the occasional artillery of rockfall. The snow softened just an hour after sunrise, so we reached the windy notch of Il Colletto and slipped back into the shade. As we began the long descent I felt a tinge of melancholy. It was an incongruous but familiar sensation, nostalgia for the present moment. In the happiest instances, like exploring a new landscape and sharing it with friends, these fleeting peak experiences bring my attention to the brevity of our little lives. Perhaps I would never see this valley again.

A combe of cold, sheltered snow led us down towards the treeline. Where the snow thinned, we hopped between patches of névé, skipping over fins of abrasive rock. Finally, we joined a 4×4 track through the larch forest and slid serenely down the last bends to the village of Saretto.

Previous spread: Finlay on the W ridge of Tête de Moïse.
Top left: Valentine and Finlay on the summit ridge of Bric Content. Bottom left: Finlay skiing a gully on the northwest flank of Bric Content.

01 Bivacco Valmaggia (2335m)

Road end (1812m) above Rifugio di Viviere, Valle Maira, Italy

12 places (12 mattresses, 14 blankets)

Gas stove in situ

Spring 60m to W

Electric lights & heating 'air pump'

2h00. +550m & 4km from Rifugio di Viviere (Hiking – T1)

Drive to Rifugio di Viviere and park just beyond this. (In winter and spring the road is usually closed at Chialvetta, 1475m.) From here, hike WNW up the Valle Enchiausa to reach Bivacco Valmaggia.

4-Day Ski Tour / Hike above Valle Maira – 52km, +4400m (Ski – 4.1)

This region lends itself to multi-day ski tours and hikes, linking huts and bivouacs along the Franco-Italian border. The terrain offers mellow skiing and well-marked trails, while also tempting you with steeper cols and peaks when conditions allow. A suggested itinerary, suitable in both directions for summer hiking and spring ski touring, is as follows: Day 1 – From Rifugio Campo Base, ascend Vallone dell'Infernetto and cross Col dell'Infernetto to reach Bivacco Barenghi. Stay here or continue over Col de la Gypière to Refuge du Chambeyron. Day 2 – From Refuge du Chambeyron, head S over Pas de la Couleta, then E over Col de Stroppia. Ascend S over Col de Portiola and E over Col de Sautron to Bivacco Danilo Sartore. Either stay here or press on over Passo della Cavalla to Bivacco Bonelli. Day 3 – Ascend SW over Col du Boeuf, then SE under Tête de Moïse / Monte Oronaye, with an optional ascent via the S couloir. Descend steeply over Col de Feuillas (40°-45°) to reach Bivacco Valmaggia. Another rewarding objective above this hut is Bric Content (2733m), accessible in both summer and winter via a short scramble (F, II-) from its SW col (2657m). Day 4 – Head E, then NW over Il Colletto (W of Monte Freid) and enjoy a long descent NW to Lago Visaisa, then N to reach Saretto, a short hitchhike from your starting point.

Tête de Moïse (3104m) via S Couloir + W Ridge (PD, III)

This provides an easy but exposed ridge scramble on interactive limestone. From the bivouac, ascend W to Col de Feuillas (2754m). Ascend NW and climb the S (actually SSE) snow/rubble couloir reaching the ridge 100m W of Tête de Moïse. Head E along the ridge following red paint marks, starting via a slanting corner on the S side (II+). Continue past small towers before climbing a 15m wall on the crest (III). Scramble along the ridge to the cross and on to the true summit (3hrs from Bivacco Valmaggia). Descend by the same itinerary, possibly making 2x15-20m rappels.

Tête de Moïse via SE Face – Via Ferrata Degli Alpini (AD-, IV)

Tête de Moïse can be climbed via the SE face of its foresummit, Cima Dronero. The route follows the spectacular military Via Ferrata Degli Alpini, built in the 1930s for observation purposes and re-equipped in 2010 (PD, III). The route starts behind Bivacco Enrico e Mario, below Col de Feuillas, and follows equipped sections past two military lookout posts/bivouacs built into the rock face, leading to Cima Dronero (3050m). From here, either redescend or continue along the E ridge of Tête de Moïse (AD-), via sections of III/IV and short rappels (max 20m, bolt anchors guide the way).

Bivacco Valmaggia is perched at the head of Valle Enchiausa, a remote side valley of Valle Maira in southwest Piemonte. The road up this valley is a dead end, and like much of Piemonte, tourist infrastructure in Valle Maira has a light footprint. Compared to Valle Varaita to the north, and Valle Stura di Demonte to the south – both of which have road cols connecting with France – the upper reaches of Valle Maira often seem deserted.

Owned by a subsection of the CAI, the hut was inaugurated in 2013 in honour of engineer Angelo Valmaggia, a prominent civil figure and local mountaineer. The hut is a square wooden structure featuring large bunk bed platforms and a small kitchen area in the corner, equipped with a gas stove.

While its size and architecture are rather conventional, this bivouac is one of the most comfortable we've encountered in winter, thanks to its ingenious heat pump. On the south side of the building, a frame of boulders is encased behind glass, functioning as a thermal storage heater. This clever design captures and stores heat, which is then released into the bivouac through vents in the wall.

Above and right: Valentine and Finlay at Bivacco Valmaggia.

The bivouac sits beneath the craggy east side of Tête de Moïse / Monte Oronaye, on the French border. This mountain, composed of friable dolomitic limestone from the Middle Triassic, still offers worthwhile climbing routes and steep ski descents. Oronaye is a prominent landmark in the region and is the most southerly 3000m peak of the Cottian Alps, the mountain range running from Col de Larche to Col du Mont-Cenis.

The frontier area is dotted with military constructions from the late 19th and early 20th centuries – keep your eyes peeled for walls, bunkers, doors and windows in the most improbable places. Three hundred metres above Valmaggia lies Bivacco Enrico e Mario, an old military barracks that, while open, is usually cold and damp inside so not particularly recommended.

Left: Valentine skiing a gully on the northwest flank of Bric Content. Above: the L line shows the S couloir of Tête de Moïse. The R line marks the Via Ferrata Degli Alpini on the SE face of Cima Dronero. The route begins at Bivacco Enrico e Mario (lower circle) and passes two other lookout posts and bivouacs (circled) high on the face.

02 Bivacco Bonelli (2330m)

Sorgenti del Maira (1630m), Saretto, Valle Maira, Cuneo, Italy

8 places (8 mattresses, 18 blankets)

Gas stove in situ

700m W, stream on the N side of Lago delle Munie

Solar panels power electric lights

2h30. +700m & 4km from Sorgenti del Maira (Hiking – T2)

From Saretto, follow the road S. In summer, parking is available at the Sorgenti del Maira campsite (1630m). From there, follow marked trails W, then S, to reach Bivacco Bonelli.

Tour of Tête de Moïse / Monte Oronaye (Hiking – T3)

This tour makes a long day hike through wild terrain (15km, +1100m). From Bivacco Bonelli, head SE up the steep scree path to Col d'Enchiausa and descend to Bivacco Valmaggia. From there, ascend W, steeply over Col de Feuillas (2754m). Descend to Lac de l'Oronaye, then continue NW and NE to reach Col des Monges. From here, descend E then SE to return to Bivacco Bonelli. This route also makes a great ski touring loop following the same itinerary (ski grade 4.1). While it is easier clockwise, with Col de Feuillas ascended on foot, skiing conditions are often better in the anticlockwise direction.

Tête de Moïse (3104m) via NE Couloir + W Ridge (PD+, III)

The NE couloir is best ascended or skied (ski grade 5.2) in spring when well filled with snow. From the bivouac, ascend SE towards Col d'Enchiausa. Below the N face of Tête de Moïse locate the base of the NE couloir – two slanting 400m gullies heading L to R to a notch on the summit ridge, 100m R (W) of Tête de Moïse. Ascend the wide lower snow gully, then continue up the L of the parallel gullies via a narrow passage (45-55°), which may have steps of ice or rock. Beyond this, the terrain eases to the notch, which also marks the top of the S couloir. From here, one can follow the W ridge to the summit of Tête de Moïse as described in the Bivacco Valmaggia chapter (3hrs from Bivacco Bonelli). On descent, return either via the NE couloir (there are several bolted rappel anchors in situ) or the S couloir.

This comfortable, well-insulated wooden cabin is situated in a stunning location high above Valle Maira. Although it has the unassuming appearance of a garden shed from the outside, the compact interior is well appointed. Two sleeping platforms line the back wall, while a narrow kitchen area to the side features a three-burner gas stove, cleverly concealed beneath the work surface.

The hut sits on the rim of a limestone cirque overlooking the azure waters of Lago d'Apsoi, beneath the summit of Tête de Moïse / Monte Oronaye. The mountain's French name means "Moses' Head", perhaps referencing the distinctive profile of the summit ridge.

Right: Valentine, Ben and Finlay skinning to Bivacco Bonelli with Monte Sautron behind.
Above and opposite: Valentine, Ben and Finlay at Bivacco Bonelli.

Tête de Moïse / Monte Oronaye (3104m)

Bivacco Bonelli serves as an ideal staging point for multi-day hikes or ski tours along the Franco-Italian border between the Aiguille de Chambeyron and Tête de Moïse. Until spring, the area remains largely deserted. Indeed, at the winter solstice, the hut receives just 10 minutes of sun per day!

Opened in 1987, the bivouac is owned by the Municipality of Acceglio. Before 2020, it was locked in summer, with the keys available from the bar in Acceglio. It is now dedicated to Piedmontese mountaineer Roberto Bonelli, a leading rock climber in the 1970s.

Bonelli died in 2016 at the age of 62, slipping on the descent after a climb in the Ailefroide valley of the Écrins. Among the many bold free-climbing routes he pioneered, he is perhaps best remembered for the first ascent of Fessura della Disperazione in Valle dell'Orco, 100km to the north.

In 1974, it was a cutting-edge achievement – 15-metre runouts protected by dubious wooden wedges made it, as the name suggests, rather desperate. With the advent of wide cams, the route has become safer and somewhat less intimidating.

Remarkably, Bonelli had only started climbing the previous year. "I had never been to Valle dell'Orco before," he later recalled, "the Fessura della Disperazione was my first climb on granite."

Above: Finlay on the W ridge of Tête de Moïse. Top left: the N face of Tête de Moïse / Monte Oronaye at sunset from Monte Chersogno. The red line shows the NE couloir and W ridge. Bottom left: Valentine below the N face of Tête de Moïse / Monte Oronaye, with the red line marking the NE couloir and W ridge.

03 Bivacco Danilo Sartore (2440m)

Saretto (1525m), Valle Maira, Cuneo, Italy

8 places (6 mattresses, 8 blankets)

No fireplace or permanent stove

Melt snow in spring. "Sorgente del Baciasset" (2186m), 700m E of hut

Solar panels power electric lights

4h30. +920m & 5.5km from Saretto (Hiking – T2)

From Saretto, follow path S16 heading up W through larch forest. Pass Grange Pausa (2056m) and continue ascending to Bivacco Sartore.

Monte Sautron / Tête de Sautron – 12km, +1100m (Hiking – T3)

This circuit, which includes an ascent of Monte Sautron, traverses a rugged landscape dotted with historical military infrastructure. Though not the highest peak in Valle Maira, its unmistakable triangular shape dominates views along the upper valley. Begin by heading W and over Col de Sautron into France, then continue W and N to Col de Portiola. To ascend Monte Sautron, follow the path up the scree slopes of the NW flank and along the final ridge to the summit. Descend back to Col de Portiola and head N to 2700m, then E and SE to La Forcellina. From there, follow the path S to complete the loop. This hike can be done in both directions and also serves as a route to reach Bivacco Barenghi.

Monte Sautron (3165m) Ski Tour – 12km, +1100m (Ski – 3.1)

Despite its steep and rugged appearance, Monte Sautron can be ascended on skis via the NW face, with only brief 30° sections. The route described above makes a great ski tour, often feasible into late spring (with an approach on foot), as the shaded NW face retains snow into summer. The steepest section of the loop is the 200m, 35° descent on La Forcellina's east side.

This exquisite triangular wooden hut, with its distinctive red roof, was built in 2011 in memory of Danilo Sartore, from Santa Vittoria d'Alba near Turin. During the Second World War, Danilo's grandfather, Giovan Battista Sartore, spent seven years stationed in these mountains, serving as a frontier guard in the military based in Acceglio.

In 1998, following excellent results in youth biathlon, Danilo enlisted in the Italian Army at the Centro Sportivo in Courmayeur. He competed for four years with the Army as part of the FISI National Youth Biathlon Team, taking part in Italian and European Cup races. In 2003, he earned his diploma as a military mountaineering instructor. That same year, while rock climbing in Saint-Pierre in Valle d'Aosta, Danilo fell and suffered fatal injuries. He was just 22.

Bruno Bonino, president of the CAI Alba section, explained: "After Danilo's death, it was his father, Teresio, who built the hut from wood, in one of the most beautiful spots in the Valle Maira. He and his wife Giovanna entrusted it to us to ensure that, even when they are no longer around, someone will keep it accessible. It can accommodate about ten people and is always open."

Teresio and his friends first built the well-insulated hut in the valley, then dismantled it, had it flown up by helicopter, and reassembled it on the current site. Made of fir and larch, the structure is clad in painted sheet metal. The compact interior is equipped with electric lights, tables and chairs. Narrow fold-down beds, suspended by chains, line the walls, while two additional sleeping spaces are tucked away in the cosy mezzanine above. Like Bivacco Valmaggia, the exterior features a case of stones encased behind glass, designed to warm and circulate hot air into the hut.

The bivouac stands on the edge of a broad plateau below Monte Sautron, amid the ruins of military fortifications built in the years leading up to the Second World War. Perched high above Valle Maira, it offers sweeping views eastward to the Po plains, perfect for catching the first rays of sunshine with your morning coffee!

Above: buried signpost in Valle di Stroppia. Right: Valentine, Ben and Finlay at Bivacco Sartore.

04 Bivacco Barenghi (2815m)

Rifugio Campo Base (1650m), Chiappera, Valle Maira, Cuneo, Italy

9 places (9 mattresses, 18 blankets)

No fireplace or permanent stove

Melt snow or filter water from Lago Vallonasso

4h00. +1200m & 8.5km from Rifugio Campo Base (Hiking – T2)

From Rifugio Campo Base, follow the S18 path NW then W, passing sections carved into the mountainside during WWII, to reach Rifugio Stroppia – a small barracks built in 1933, now a locked bivouac. Continue WNW up Montagna di Stroppia towards Brec de Chambeyron, then turn N to reach the bivouac. In winter, the section below Rifugio Stroppia can be steep and icy, making it better to ascend NNW up Valle dell'Infernetto, then SW over Colle dell'Infernetto, past Lago della Finestra, and NNW to the bivouac.

Tour of Brec de Chambeyron – 12km, +1000m (Hiking – T3)

This tour traverses wild, mineral landscapes and makes a fine loop in either direction. From the bivouac, head S then SW to Col de Stroppia, descend W then NW, and ascend over Pas de la Couleta. Pass by Refuge du Chambeyron or continue NE directly to Col de la Gypière, just above the bivouac. Save some energy for the final ascent of Tête de la Fréma (3142m), a straightforward summit with a sublime panorama. This route also makes an excellent, non-technical ski tour (ski grade 2.3).

Aiguille de Chambeyron (3412m) – SE Ridge (400m, AD+, IV+)

This is the highest summit in the Alps south of Monviso. While much of the mountain consists of loose rock, the upper SE ridge is formed of colourful, compact limestone of excellent quality. The normal route can be climbed year-round but is best in late spring when consolidated névé covers the loose rock. Graded PD (II+), it ascends the S face via the central Couloir Gastaldi (a fine ski line when conditions align, 45-55°, ski grade 5.3) then traverses R (E) along ledges (two options) to reach the SE ridge at Brèche Nérot-Vernet. Both the normal route and SE ridge route follow the crest, briefly traversing on the R (NE) side to reach a notch in the ridge leading to the Aiguille Orientale. The normal route then traverses back L (W) to rejoin Couloir Gastaldi and ascend the upper S face to the principal summit. A finer option, on better rock, is to climb the SE ridge over the Aiguille Orientale (IV+). For this, leave the normal route and scramble up to reach a pink slab at the base of the Aiguille Orientale. Climb the slab, then a ramp on the R leading to an awkward chimney. Climb this and continue up to the Aiguille Orientale. From here, scramble down along the crest, bypassing a gendarme on the N side to reach a notch, then climb up to the main summit. The descent via the normal route is delicate and exposed.

Brec de Chambeyron (3389m) – North Couloir (450m, D-, 55°)

The striking tower of Brec de Chambeyron forms the northernmost point of a precipitous 3km wall that marks the Franco-Italian frontier. The N couloir is best climbed in late spring when well filled with consolidated snow. From the foot of the N face, ascend the wide entrance gully shared with the Couloir Bujon. After 100m, fork R to follow the N couloir, climbing to the summit plateau before heading SE to the summit. For the descent, either return via the N couloir – an engaging steep ski (5.4) – or via the adjacent Couloir Bujon or the normal route. To descend via the normal route (AD-), head 30m N from the summit, rappel a chimney and then descend ENE to a shoulder at 3320m. From here, the normal route zigzags NE down ledges and gullies. To descend Couloir Bujon (AD, 40-50°), from the shoulder at 3320m downclimb ramps trending NNE to reach the broad gully, then descend this (ski grade 5.1) to the base.

Bivacco Barenghi, owned by the Cuneo section of the CAI, was inaugurated in 1958 in honour of Giuseppe Barenghi, a Milanese engineer and one of the most promising Italian mountaineers of the post-war era. He died in 1954, aged just 34, during Piero Ghiglione's expedition to Mount Api (7132m) in Nepal – the first Italian attempt on a peak over 7000m.

The expedition was marked by tragedy, beginning with the loss of doctor Roberto Bignami, who slipped into a river while crossing a precarious log bridge on the approach. Later, Giuseppe Barenghi and Giorgio Rosenkrantz perished after being caught in a storm trying to make the first ascent of Mount Api with Sherpa Gyaltzen Norbu.

Records differ on whether or not they reached the summit. Rosenkrantz fell ill near the summit and waited for Barenghi and Norbu, who may have reached the central summit. Barenghi then pushed on alone towards the western summit but was lost in the storm.

Above: Finlay and Valentine in Bivacco Barenghi.
Previous spread: Brec de Chambeyron's west face at sunset from the Col de Vars.
Top right: Valentine at Bivacco Barenghi with the E face of Brec de Chambeyron behind. The red line shows the normal route. Bottom right: Brec de Chambeyron's north face. The right line shows the north couloir. The left line shows Couloir Bujon.

Brec de Chambeyron (3389m)

Aiguille de Chambeyron (3412m)
Aiguille Orientale
Brèche Nérot-Vernet

Norbu descended with Rosenkrantz, who survived a first night sheltered in a crevasse at nearly 7000m but froze to death during a second night at around 6600m. Only Norbu made it down alive. Though the expedition's outcome initially sparked controversy, it was soon overshadowed by the success of the Italian first ascent of K2.

The original Bivacco Barenghi was a compact sheet-metal "Apollonio" shelter. Restored and expanded in 1984, it is now lined with wood and features a small vestibule/kitchen adjoining the original nine-berth sleeping area. If the bivouac is full, Refuge du Chambeyron, just an hour away across the French border, has a comfortable winter room with a wood-burning stove. While Bivacco Barenghi is an excellent stopover on a multi-day hike or ski tour, it's more enjoyable in summer, when the sun is higher in the sky and the shelter is less cold and damp.

Situated in one of Valle Maira's most beautiful and remote corners, on the border with the French Ubaye, the bivouac is reached via a spectacular hike that passes the Cascate di Stroppia. During the peak snowmelt in early summer, these waterfalls plunge 500m, making them Italy's tallest natural waterfalls. Though the hut's view is restricted by surrounding rock walls, a short hike (T3) up the Tête de la Fréma, on the French border, reveals a spectacular panorama across the entire region.

Above: Finlay, Ben and Valentine in Bivacco Barenghi. Top left: Finlay, Valentine and Ben ascending Vallone dell'Infernetto. Bottom left: the south face of the Aiguille de Chambeyron. To reach Brèche Nérot-Vernet, two options are shown on the topo: the lower line follows better rock (III, marked with white paint dots), the upper approach ledge is the traditional route (II+). From Brèche Nérot-Vernet, the upper line shows the SE ridge route over the Aiguille Orientale, while the lower line marks the normal route across and up the south face.

05 Bivacco di Punta Venezia (3080m)

Pian del Re (2020m), Valle Po, Italy

2 places (3 mattresses, 10 blankets)

No fireplace or permanent stove

Melt snow or collect water on approach

3h00. +1070m & 4km from Pian del Re (Hiking – T4)

From Pian del Re, follow the path S past Lago Fiorenza, then ascend W to Rifugio Giacoletti. From the refuge, take the rocky path NW to the base of Couloir du Porc, the gully descending E from the col. Climb this via the steep path equipped with chains and ropes on the R (N) flank. In early summer, this may be a steep snow gully requiring crampons. From the col, head 400m N up the S ridge of Punta Venezia to reach the bivouac, just below the summit. The final section to reach the summit cross involves a short technical climb (F, II+). There are bolt anchors for protection, and a rappel is possible for descent.

Tour of Punta Venezia – 15km +1500m (Hiking – T4)

Follow the approach described above from Pian del Re to Punta Venezia. Return to Col du Couloir du Porc, head SW, and then NW to Refuge du Viso. From there, continue N then NE to reach Col de la Traversette. Pass through the tunnel back into Italy, then head SE to complete the loop back to Pian del Re.

Monviso (3841m) – Berhault Traverse (D, V)

This complex ridge progressively increases in difficulty as it extends south toward Monviso, covering over 5km of crest and 2000m of elevation gain. For most teams, the full traverse takes 3-4 days. To begin, approach from either Italy (Pian de Re) or France (Roche Ecroulée) to gain Col de la Traversette.

***Section 1 (4-6hrs)** – From Col de la Traversette, ascend the N ridge (II) of Rocce Fourioun. Continue S near the crest over loose terrain to reach the 15m gendarme of Aiguille Bleue, which is climbed via a chimney (III+). Head S, then descend the broad SW ridge to locate an awkward gully (cairns) leading to Col du Couloir Blanc. From here it is possible to escape SW.*

Continue via Cresta Venezia (AD-, IV-), the NW ridge of Punta Venezia. Traverse under the first point, then follow the ridge to reach a steep gendarme. Rappel 15m, then continue along the fine blocky crest (IV-) to the summit cross of Punta Venezia. From the summit, either downclimb (III) or rappel 15m E to a ledge, then follow this S to Bivacco di Punta Venezia. From the hut, follow the path S to Col du Couloir du Porc. From here it is possible to escape W to Refuge du Viso or E (via ferrata) to Rifugio Giacoletti.

***Section 2 (6-10hrs)** – From Col du Couloir du Porc, ascend the NW ridge of Punta Udine. Continue c.50m S to a notch, then over the next point. Locate a bolt anchor on the E side and make 2 x 25m rappels to a ledge, then head S to reach Brèche des Anges. From here, follow the crest over 6 pinnacles, navigating numerous climbs and rappels (max. 30m) along the intricate 800m NNW ridge of Punta Roma (AD, IV).*

From Punta Roma, it is possible to escape down the normal route via the E face (F, II+). Otherwise, continue down the S ridge (AD, III+) to a vertical drop. Make a 25m rappel then continue down to a notch, Passo del Colonnello (possible escape E). Follow the ridge S over numerous points, bypassing a large gendarme on the W, before reaching Passo Giacoletti (possible escape on E and W sides).

This tiny hut, also known as Bivacco degli Amici, sits just a few metres below the summit of Punta Venezia (3095m), a popular peak on the Franco-Italian border. It lies a couple of kilometres north of Monviso, the highest peak in the Cottian Alps. Built in 1937 as a World War II Italian military observation post overlooking the high French valleys, it was part of a network of similar structures. This included the shelters at Rocce Fourioun and Colle delle Traversette, now reduced to ruins.

Standing just 1.5m high, this metal-clad, half-cylinder wooden hut resembles the earliest wooden bivouacs still in use, such as Bivacco Craveri and Bivacco Adolfo Hess in the Mont Blanc massif, which date back to the 1920s. Though it was renovated in 1991, it remains very rustic, and plans are afoot to replace it with a slightly larger, modern shelter. While the bivouac is of Italian origin, the exact location of the frontier requires more accurate measurement, as the hut may lie within French territory.

For most visitors, this hut serves as a scenic picnic halt after ascending Punta Venezia, as nearby Rifugio Giacoletti or Refuge du Viso provide rather more salubrious accommodation. However, for adventurous types, Bivacco di Punta Venezia offers a spectacular and solitary overnight perch on this jagged frontier ridge, with an unforgettable sunrise view over Monviso.

Above: Tom Coney and Teresa Schaubeck on the Punta Roma traverse.
Top right: Tom and Teresa approaching Bivacco di Punta Venezia.
Bottom right: Valentine descending towards Pian del Re.

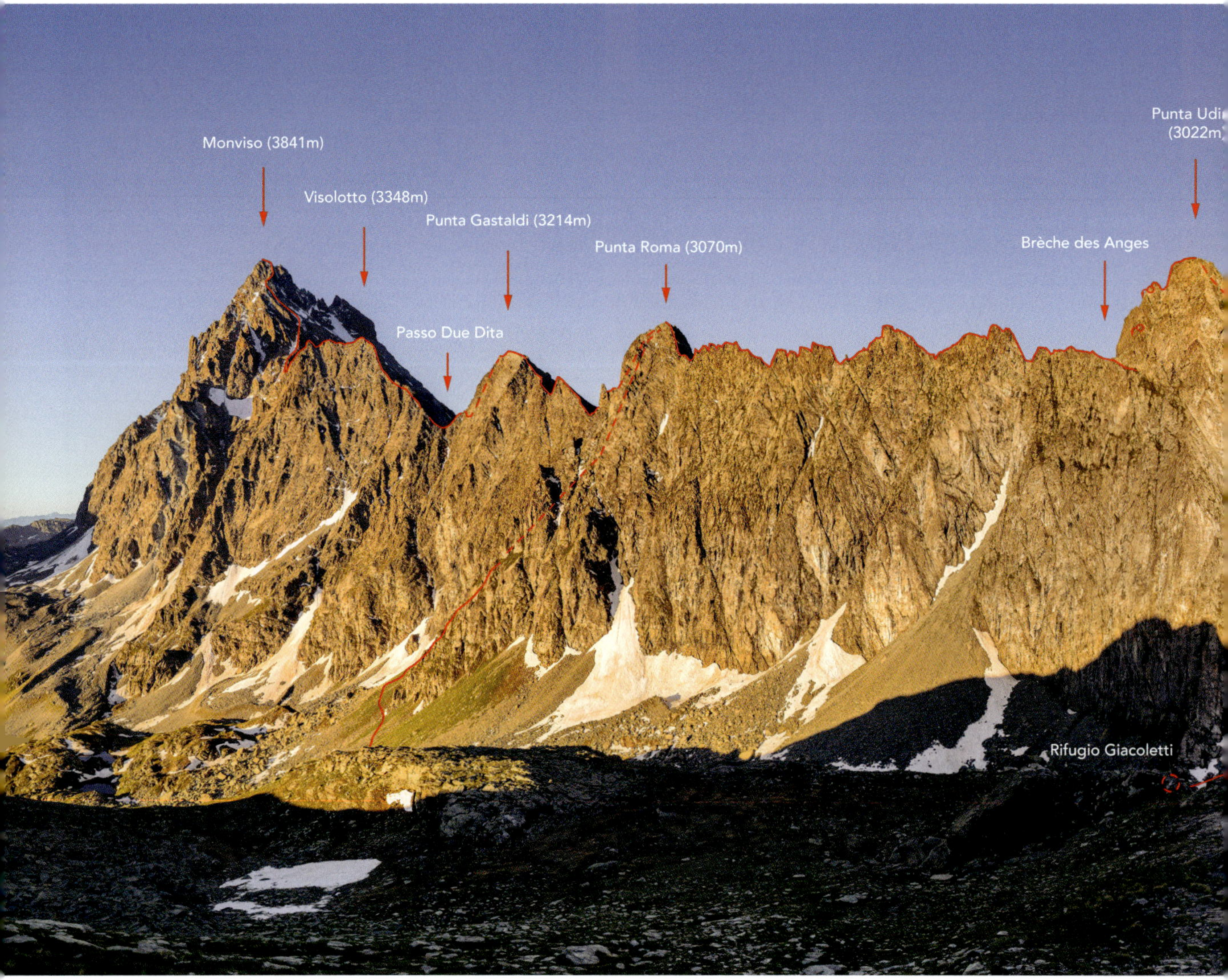

Section 3 (8-12hrs) *– From Passo Giacoletti, follow the N ridge of Punta Gastaldi (PD+, III+), climbing over a first summit before descending to a notch via a 12m rappel. Ascend a ramp, then a corner (III) to regain the crest and follow it to the summit (escape possible WSW, F+). Descend the S ridge (F+) to a notch W of a small summit, Due Dita. Climb over Due Dita, then descend or rappel a steep couloir on the L (SE) side to reach Passo Due Dita (escape possible SW).*

Ascend the NW ridge of Visolotto (AD, IV), bypassing a gendarme on the R via a narrow gully. Continue along the crest, then traverse a rising ledge on the L (NE) side to reach a notch. From here, climb (IV) on the R of the crest. Near the summit, head R (S) across grey slabs to gain the SW ridge, following it for 30m to the main summit. Continue SE (II) to the central summit, then scramble down (III) before making a 15m rappel or steep downclimb (IV) to a notch. Ascend to the east summit, then descend the SSE ridge, making 4 rappels to reach Colle Nord delle Cadreghe and traverse to Colle Sud (escape possible SW).

Section 4 (3-6hrs) *– From Colle delle Cadreghe, ascend the NW ridge of Monviso (D-, IV+). Begin by climbing a scree cone and a crumbling buttress to the base of a red pillar. From there, if snow conditions permit, ascend Couloir Perotti, the 300m, 50° gully to the L of the red pillar. Otherwise, climb directly up the N ridge of the red pillar (IV+). From the notch behind the red pillar, either follow the snow gully heading R (S) towards the W face, then ascend a snow ramp L to regain the NW ridge, or climb directly up the crest (IV+). Follow the ridge (III+), bypassing a large gendarme resembling an inverted bell on the R, then ascend the W face (III, rotten) to reach the summit of Monviso. From here, descend via the normal route on the S face (PD-, 3-4hrs).*

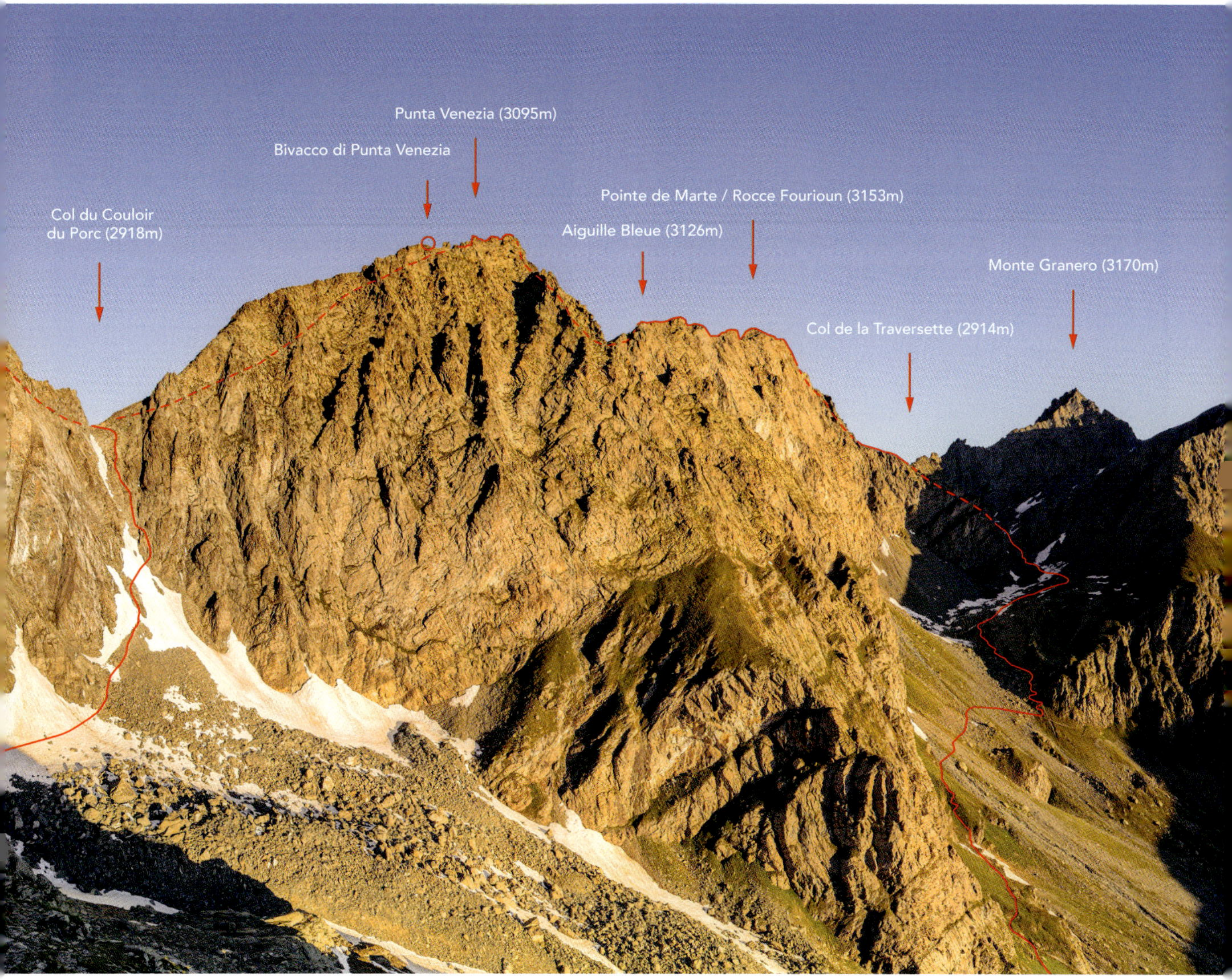

Punta Venezia was likely first ascended in the mid-19th century by military surveyors or chamois hunters. Christian Almer (Jr.) and William Coolidge discovered a cairn on the north top during their first recorded tourist ascent in August 1888. Coolidge, an Anglican priest and prolific mountaineer, made hundreds of first ascents across the Alps, including meticulous explorations of the Cottian and Dauphiné Alps in the 1870s and 80s. Though somewhat eccentric, he is remembered as one of the most fastidious alpine historians and guidebook authors. Having moved to Switzerland at 35, his obsession with historical accuracy earned him the nickname "the curmudgeon of Grindelwald" among British climbers.

While reaching Bivacco di Punta Venezia is just a technical hike, the final 20 metres to the summit require easy but exposed climbing (F, II+), now protected by bolts. Punta Venezia's precipitous east face also offers excellent multi-pitch climbing routes, just a short walk from Rifugio Giacoletti.

For alpinists, Punta Venezia is best known for its position along the great north-northwest ridge of Monviso, and the hut can make for a strategic first night on this climb. The initial section, named the Skywalker ridge in homage to Coolidge, stretches from Col de la Traversette to Col Couloir Blanc, followed by the Cresta Venezia to Punta Venezia. This section alone makes for an excellent outing in its own right.

Above: overview of Monviso's Berhault Traverse, seen from Rocce Alte above Rifugio Giacoletti.

The full enchainment, an epic journey known as the "Berhault Traverse", follows the frontier crest from Col de la Traversette to Monviso, crossing seven 3000m summits before finally reaching Monviso. Before Patrick Berhault's outing, the ridge had only been climbed in its entirety twice, both times in summer. The first complete ascent was in 1968, over four days, by Italians Hervé Tranchero and Livio Patrile. Then in 1990, Riccardo Maero made a solo traverse in just eight hours. Berhault's solo winter ascent in 2001 formed part of his Grande Traversée des Alpes, an extraordinary 167-day human-powered journey across the entire Alpine arc, from Triglav in Slovenia to reach the Mediterranean at Menton.

Between August 2000 and February 2001, he climbed 22 major summits, often with friends, but undertook long sections – including this one – alone.

Climbing in the depths of winter, he noted: "At the Col de la Traversette in the storm, it feels like Patagonia! ... It's tough, but oh, how beautiful it is! The narrow ridge, lined with enormous cornices, is besieged on its northern face by immense ice flutes. Not a soul in sight – it feels like the Cordillera Blanca or the Himalayas."

Above: Monviso and the Berhault Traverse at sunrise from Rocciamelone. Top left: Simon-Pierre Génot and Tom Coney approaching Aiguille Bleue. Top right: Mark Charlton, Stu McAleese and Miles Perkin by Bivacco di Punta Venezia. Bottom left: Tom Coney and Teresa Schaubeck on the Cresta Venezia.

06 Bivacco Tornior (2552m)

Thures (1635m), Valle di Susa, Piemonte, Italy

6 places (6 mattresses, 15 blankets)

No fireplace or permanent stove

Stream 90m N

4h00. +920m & 12km from Thures (Hiking – T1)

From Thures, follow the 4×4 track up the valley to Ponte Thuras (1945m). During the summer months, driving to this point is sometimes permitted. Continue along the same track to reach the bivouac at the head of the valley. The route is also accessible on skis during winter.

Punta Ramiere (3305m) – WNW Ridge (Hiking – T3 / Ski – 2.2)

From Bivacco Tornior, follow the path NE to Colle della Ramière, then continue ESE up the scree path to the summit, (3h30 from the bivouac). Alternatively, a loop can be made by heading ESE from Bivacco Tornior to Col des Thures, then following the SW ridge to the summit (T5, shale and several rocky steps from 3050m). The WNW ridge route also offers a fine ski tour in spring.

Bivacco Tornior is a simple, compact and solidly constructed square wooden hut. Erected in 2008 by the Tornior family in memory of their son Andrea, who perished in the mountains, it is easily accessible from the Italian Susa valley and via Col des Thures from Abriès in French Queyras.

The hut sits in the upper reaches of the Val Thuras, a deep valley surrounded by peaks of shattered rock. Punta Ramiere (or Bric Froid in French) rises to the northeast, at the head of the valley. Straddling the Franco-Italian border, it is one of the highest mountains in the Cottian Alps. On the first ascent in September 1877, made directly up the steep scree slopes of the south face, French alpinist André Salvador de Quatrefages observed: "The peak, isolated from any rival, commands an extensive panorama: opposite us, to the west, separated from us by a narrow valley, the Grand-Glaiza, like a ruined wall; in the distance, Mont-Rose, the Matterhorn, the superb Weisshorn, Mont Blanc and the Pelvoux massif." Having been chased from the summit by threatening clouds, his party descended swiftly, glissading down the scree. He concludes, without a hint of snobbery, "the ascent of Bric-Froid is extremely easy; even the most inexperienced tourist should hardly be put off."

Top: David Lee hiking up Punta Ramiere. Centre: David Lee and Judith Forget-Piedalue poring over the map. Right: an unexpected encounter with Scottish Highland cattle.
Opposite page: Bivacco Tornior at sunrise.

Now, as then, Val Thuras offers rich mountain pasturage. Expecting to encounter local breeds of sheep or cattle, we were surprised to pass a large herd of Scottish Highland cattle, and then higher up the valley, a drove of 30 donkeys. At its peak in the 19th century, the hamlet of Thuras comprised 80 buildings and supported a thriving agricultural community. The 18th-century Chapel of San Rocco is the only original structure that survives from that era.

The soft, grass-covered landscape scarcely conceals a sombre past. Hidden throughout the region are extensive military bunkers and gun installations, dating from the 19th century through to the Second World War.

In the 1930s, the Italian army constructed a road stretching all the way up Val Thuras, and a thread of rustic telegraph poles still marks the lines of military communication that once threaded these border valleys. Just 300m south of Bivacco Tornior, up on the hillside, stands the concrete shell of the large "XVI" barracks. Much of the frontier crest is still strewn with braids of barbed wire and the ruins of lookout posts.

In the shattered "XXI" barracks near Colle della Ramière, we ran into a large male ibex sheltering from the sun. He was just as surprised as we were to meet in this austere bunker. In 1940, this valley served as an access point during Italy's short-lived invasion of France in the first "Battle of the Alps". Later in the war, after Italy had switched sides to join the Allies, German forces burned the village of Thuras to the ground in 1944, in reprisal for an attack by an Italian resistance group.

Above: Bivacco Tornior at dusk. Top right: Adam Bramley and Valentina Eula heading to Col des Thures.
Right: David passing barbed wire along the frontier ridge. Far right: David and Judith in the entrance of Bivacco Tornior.

VIETATO
FUMARE

With David and Judith, friends visiting from Canada, we had driven three hours in the hope of visiting the novel architectural object that is Bivacco Corradini. On arrival at the trailhead, we found a discreet notice announcing that it was closed for maintenance. It was already late in the afternoon, so with a quick change of plan, we set out for the nearby Bivacco Tornior. Shortly after setting off, the sky darkened and thunder grumbled at the far end of the valley. Fog drifted in, carrying a veil of light rain. By dusk, after a few hours of brisk marching, a log cabin emerged through the mist.

Circling around to the front, we found the shutters removed and a glimmer of light spilling onto the grass. Opening the door, we startled a half-dressed girl who clearly wasn't expecting visitors so late in the day. Valentina and her partner Adam had a family house in the nearby town of Oulx, and they too had hoped to visit Bivacco Corradini.

Bivacco Tornior was far less widely known, so stumbling upon such a cosy hut, deep in the mountains, felt like an unexpected success for all of us and we spent the evening chatting about other hidden corners of the region we had each explored.

After dark, the sky cleared and we stood outside enjoying the warm evening air of late summer and the rich, earthy scent of the moistened ground. Behind us, a silent electric battle sparked and flickered across the sky somewhere beyond the French border, casting jagged shadows of the "ruined wall" of Grand Glaiza across the hillside. I wondered whether the soldiers once stationed at the "XVI" barracks beneath that wall had ever appreciated the scent or the sight of this mineral landscape, despite the horrifying context of war.

Top left: on the summit of Punta Ramiere / Bric Froid. Bottom left: David brewing the morning coffee. Above: David and Judith heading down Val Thuras.

07 Bivacco Lago Laus (2274m)

Borgata Seleiraut (1530m), Val Chisone, Piemonte, Italy

4 places (4 mattresses, 4 blankets)

Wood stove present, but fuel supply may be limited. No trees nearby

Spring fed through a plastic pipe, 100m NE on the lake shore

2h30. +780m & 5km from Borgata Seleiraut (Hiking – T2)

From Borgata Seleiraut, follow the path E then N past Truc del Cuculo to reach Lago Laus. Alternatively, between June and October, a dirt road provides vehicle access to Rifugio Selleries, reducing the approach to the bivouac to 1hr. To reach the Rifugio, leave the main Val Chisone road at Depot, ascend the road N to Prato Catinat, then at 1768m follow the track 5km E. From Rifugio Selleries, a trail leads E to Bivacco Lago Laus. Both approaches are possible in winter on skis.

Giro dell'Orsiera – 55km +4400m (Hiking – T3)

This trek makes a loop around the region through varied terrain in forests, alpages and across rugged mountain cols. It is traditionally divided up into six short stages from refuge to refuge, but a fit team could easily complete it in three days.

Punta Cristalliera (2802m) via North Flank (Hiking – T3)

From Bivacco Lago Laus, head NE past Lago della Manica to reach Colle Superiore di Malanotte. Head SSE, following cairns through blocky terrain to reach the summit cross (2h from Bivacco Lago Laus).

Bivacco Lago Laus occupies a large room in one of the park service's pastoral huts on the edge of the lake. Renovated in 2015, these rustic stone buildings have thick walls, but the bivouac has a concrete floor and limited insulation. As such, it stays cool in summer but is likely to feel damp in winter. The room includes four beds, a dining table and a good wood stove. There is electric lighting and, depending on availability, firewood may be stocked in the adjacent shed.

Punta Cristalliera, the prominent peak towering over Lago Laus, is a classic and popular hiking objective. One of the highest summits in the Orsiera Rocciavrè Nature Park, it offers a fine panorama. An early start is recommended as clouds tend to build by late morning and often obscure the views.

The Parco Naturale Orsiera Rocciavrè spans much of the mountainous region between the Valle di Susa to the north and the Val Chisone to the south. Covering over 10,000 hectares, the park centres on the main ridge stretching from Monte Orsiera over Punta Cristalliera to Monte Rocciavrè. The mid-range altitude of these peaks, along with mild temperatures and low rainfall, gives the area a particularly long hiking season. A beautiful multi-day route, the Giro dell'Orsiera, forms a 55km circuit of the park and traverses habitats rich in wildlife, including ptarmigan, boar, mouflon, wolves, chamois and ibex.

Below: Bivacco Lago Laus. Right: hut approach and interior.

With three of us in my tiny, underpowered car, we only just managed to crawl up the steep thread of hairpin bends to the hamlet of Seleiraut. On the woodland's edge above the cluster of stone buildings, we scavenged the last of the raspberries bristling along the path. After a couple of hours of hiking uphill, we passed Rifugio Selleries, which I had assumed was a remote mountain refuge. The sight of twenty parked cars quickly dispelled that illusion. Despite standing at over 2000m, you can still drive right to the door.

I had lived with David for several years while we were at university in Scotland, and we'd shared plenty of mountain adventures. His partner Judith, however, had grown up in the swamps of Nova Scotia and wasn't sure how she'd fare in steep alpine terrain. She'd warned us that she suffered from severe vertigo, so I had chosen a route without technical sections or exposure. My original plan had been to continue up to the tiny chapel-bivouac on Mont Robinet, but this was quickly revised when even a level stretch of path with a steep drop on one side made her dizzy.

After that unsettling section, we paused to chat with some ibex grazing beside the path, and the distraction helped ease Judith's rising tension. Still, the sloping trail to Bivacco Lago Laus proved to be a genuine challenge. I had never witnessed anyone with such extreme vertigo.

The lake was surrounded by a herd of 30 cattle that were curious and skittish in equal measure. On the slopes above, we befriended a pair of donkeys who later wandered down to the hut and stood inquisitively at the door for much of the afternoon. Storm clouds gathered at sunset, but released only a few spots of rain. As dusk fell, the brooding clouds drifted away and lit up the distant sky with a magnificent lightning display. The constant chime of cowbells, however, made sleep all but impossible.

Left: Bivacco Lago Laus with Punta Cristalliera behind. Above: Judith with the local donkeys.

08 Bivacco Robinet (2679m)

Borgata Seleiraut (1530m), Val Chisone, Piemonte, Italy

5 places (10 mattresses, 20 blankets)

Gas stove in situ

Weak springs, 300m W of Colletto Robinet

4h00. +1200m & 7km from Borgata Seleiraut (Hiking – T3)

From the end of the steep road at Borgata Seleiraut, follow the path E then N past Truc del Cuculo. Continue N around Monte Glantin, then NE (past marked springs) to Colletto Robinet. From there, head ESE to reach Monte Robinet and the Cappella Madonna degli Angeli, which houses Bivacco Robinet.

5h00. +1630m & 9km from Case Agostino (Ski – 2.2 / Hiking – T3)

Monte Robinet and Monte Rocciavrè are classic ski tours from the Vallone della Balma. As the route begins and ends at a relatively low altitude, it's best suited to midwinter, when the snowline is low. In summer, the same approach is also popular on foot. From Case Agostino (1150m) above Coazze, follow the valley WNW to Rifugio Balma. Continue W to Colletto Robinet, then ESE to Monte Robinet.

Monte Rocciavrè (2778m) via ESE Ridge (Hiking – T4)

From Monte Robinet, descend NNW to Colletto Robinet. Head N up the blunt ridge, then scramble WNW (cables) to reach the summit of Monte Rocciavrè (40mins from Bivacco Robinet).

Bivacco Robinet is notable both for its dramatic location, perched on the summit of Monte Robinet high above the Po plains, and for its unusual setting within the Cappella Madonna degli Angeli, a high alpine chapel. Monte Robinet stands at the head of Val Sangone, near the watershed dividing the major Piemontese basins of Val Chisone and Valle di Susa. On clear mornings, the views are exceptional, sweeping from Monviso in the south to the Écrins in the west and Monte Rosa in the north. Often in these mountains, cloud or fog swiftly engulfs the summits in the morning.

Thanks to its easy access and panoramic views, Monte Robinet has long been a popular destination for hikers. The chapel was originally built in 1900, but over the decades maintenance was intermittent. By the early 1970s, members of the local alpine club recognised the need for urgent repairs. Volunteers from the Coazze section of the CAI took on the task, consolidating the foundations and walls before constructing a Faraday cage to better shield the structure from lightning strikes. The southern aspect of the chapel was clad in steel, giving the building a unique appearance, resembling some ancient spacecraft. In recent years, solar panels and electric lighting have also been added.

Below: Cappella Madonna degli Angeli at sunrise. Top right: David inside Bivacco Robinet. Bottom right: David brewing coffee on the steps of Cappella Madonna degli Angeli.

MUSTO

The bivouac facility is a small annexe attached to the east side of the chapel, completed in 1975 with the construction of the sleeping platform. As this annexe provides direct access to the chapel interior, this room was likely originally intended to be a sacristy. The compact space contains a table, benches and two single bunk beds, with a low wooden mezzanine above holding three more mattresses. With rendered stone walls and a concrete floor, Bivacco Robinet is most suited to summer use, when the cool interior offers relief from the sweltering plains below.

The chapel itself is spacious, featuring a wooden floor and a massive stone altar. To one side, five extra mattresses are stacked for visitors should the bivouac be full. The walls are adorned with colourful religious murals and countless memorials and votive offerings, creating a striking contrast with the mineral tones of the surrounding landscape.

Top left: Cappella Madonna degli Angeli on Monte Robinet. Top right: the altar inside Cappella Madonna degli Angeli. Bottom left: interior of the chapel. Bottom right: David at the summit of Monte Rocciavrè. Opposite: Cappella Madonna degli Angeli, with Monviso in the distance.

09 Bivacco Berrone (2860m)

Beaume (1098m), Oulx, Italy

8 places (4 double mattresses, 3 blankets)

No fireplace or permanent stove

Melt snow or filter water from nearby lakes

5h30. +1780m & 9km from Beaume (Hiking – T2)

From Beaume, hike N up the spur on the W side of Vallone del Seguret to reach Serre du Quin and the military road at 2226m. In summer, this gravel track can be accessed from Salbertrand by mountain bike, and on two days a week by motor vehicles, with parking available at Colle Pramand. Follow the path N to the ruins of the Vin Vert barracks, then continue NE to reach Bivacco Berrone by Laghetti del Vallonetto. A short hike beyond leads to Cima del Vallonetto (3217m), the peak at the head of the valley.

Traverse of Massif du Mont-Cenis – 36km, +2700m (Ski – 3.3)

This region offers numerous possibilities for hiking and ski-touring traverses. Bivacco Berrone is ideally placed as an overnight shelter for a winter traverse from Rochemolles (Bardonecchia) to Moncenisio (Susa). From Rochemolles, ascend SE into Valfredda, then climb over Colletto Vin Vert and head NE to reach Bivacco Berrone. The following morning, either ascend to Cima del Vallonetto and scramble along the ridge to reach the Galambra plateau, or ski NW into upper Valfredda and ascend to Passo Galambra. From there, climb to Pointe Sommeiller (3332m), then ski NE along the frontier and descend Glacier d'Ambin (45°). Continue ESE to Col d'Ambin, then scramble E up the ridge and head NE to Col de l'Agnel Supérieur. Descend NE to reach Bivacco Hannibal for a second night. On the third day, continue NW past Lac de Savine and ascend to Col des Lacs Giaset, then descend NE to reach the road leading to Moncenisio. A similar high-level traverse is possible on foot, although from Passo Galambra it is easier to head first to Bivacco Mario Sigot, then follow the path NE across the hillside to Col d'Ambin, where you can rejoin the route described above.

With its whimsical and formalist design, Bivacco Berrone, known as the "pinwheel shelter", stands as one of the most architecturally adventurous alpine huts to date. Shaped like a children's paper fortune teller or an intricately cut jewel, it tests the boundary between sculpture and shelter. Designed by architects Andrea Cassi and Michele Versaci, who also created the nearby Bivacco Corradini, it was conceived as a "rigid tent" in response to a design brief to "install a light, minimally invasive and totally reversible structure". The result pushes the notion of habitable architecture deep into the realm of art.

The hut is located in the upper Rio Seguret basin, beside the Laghetti del Vallonetto, in a bare, mineral landscape high above Valle di Susa. Built in 2023, it honours Stefano Berrone, an alpinist from Sauze d'Oulx who died in a paragliding-speedriding accident the previous year at the age of 40. The land belongs to the municipality of Oulx, while the project was conceived and funded by Berrone's family and friends.

Above: Valentine and Misha in Valfredda. Right: Bivacco Berrone at sunrise.
Opposite top: Bivacco Berrone at sunset. Opposite bottom: inside Bivacco Berrone.

Perched on a small gravel rise, the bivouac is elevated on four concrete plinths allowing wind to rush underneath. This helps prevent snow accumulation and insulates the shelter from ground moisture. Constructed from slabs of cross-laminated timber (CLT) fixed to a steel subframe, it was designed to be lightweight in both form and environmental footprint. Prefabricated off-site, the hut was flown up in sections by helicopter and can easily be dismantled and removed if required.

Folded into 16 facets, the angular design resembles a piece of origami, playing on the visual tension between rigidity and pliability. For such a small interior volume, the surface area is remarkably high. Even so, the shelter is well insulated: on a cold winter evening it warmed quickly from body heat and a cooking stove, and retained warmth well through the night. Externally clad in aluminium and punctuated by bold triangular windows, its crystalline form stands in striking contrast to the drifting snow and sinuous scree slopes. "Engaging in a dialogue with the alpine light, the colours of the rocks, and the surrounding snow slopes," explains Versaci, "it creates a constant interplay of reflections and responds dynamically to changing weather conditions."

Inside, the compact scale and sloping roof panels compel visitors to stoop, crawl and sit, creating an intimate, tent-like atmosphere that is reinforced by the red-painted wood. The hut is laid out in a radial plan: four wings extend from a symbolic central "hearth", with seating on storage trunks recessed into the floor. These contain pots, pans and a portable gas stove, but as nothing is fixed, it's advisable to bring your own. The wooden sleeping platform accommodates up to eight people on four large green suede mattresses.

The central window faces out across the Susa valley towards the strange, flat-topped silhouette of Mont Chaberton (3131m), crowned by eight curiously geometric forms. Between 1898 and 1910, Italian troops built an artillery battery here, still the highest in Europe. They excavated six metres from the summit to level it, then installed eight turrets, each twelve metres high. At the time, the fort's exposed gun towers were considered unreachable by enemy fire. However, cannon technology advanced and during the Second World War on 21 June 1940, French forces destroyed six of the turrets using 280mm howitzers from concealed positions 10km away above Briançon.

Above: Misha brewing tea. Left: Bivacco Berrone at dusk. Previous spread: Valentine and Misha skiing up Valfredda.

In 2025, winter had been slow to arrive in Piemonte, but in early March a heavy snowfall blanketed the region, pushing the snowline to the valley floor. Setting off from Rochemolles under a timid sun, Valentine, Misha and I skinned past Mouchecuite, a rustic cluster of summer dwellings still buried in drifts, then continued up the larch-lined gorge leading into Valfredda. The narrow valley gradually opened into a broad, flat plain, where we paused for a late lunch on the warm stone steps of the summer refuge. Far above us, three ski tourers were approaching the fortifications of Monte Jafferau on the skyline. Built in the 1890s, just before the summit fort on Mont Chaberton, this artillery battery formed another key element in the defences of the Bardonecchia basin. Apart from their distant figures, the valley lay silent, marked only by the tracks of hare and ibex.

As we ascended, a northerly breeze swept over the ridgeline. The light grew flat, the snow reflecting the muted sky. Opposite us, a dozen chamois grazed peacefully on a grassy spur where the snow had already receded. Upon reaching the crest near Colletto Vin Vert, we spotted the strange, faceted form of the bivouac in the distance. We skied into the valley below, and as we approached, the shelter re-emerged from the horizon like a grounded spacecraft in the late afternoon light.

This hut provides a strategic overnight point in this area, especially in winter, as the nearby Bivacco Vallonetto is ill-suited to cold conditions and the refuge in Valfredda remains closed. Yet for all its visual charm, the Bivacco Berrone has its quirks. The tiny triangular doors are awkward to crawl through, and it's difficult to avoid dragging snow inside. The design often prioritises form over function. It's playful, if occasionally impractical.

After a cup of tea, I stepped back outside to take photos as a glimmer of sunset pierced the blanket of clouds to the west. My hands soon grew numb, but I lingered until dusk, the frost searing my fingers. Later, lying in bed, I gazed out over the valley, where streetlights flickered beneath a canopy of stars. While Valentine and Misha slept, I turned over and rested my head against the glass, watching planes drift silently overhead and constellations wheel across the night sky.

Top left: Valentine and Misha skiing up Valfredda. Top right: Misha squeezing through the entrance of Bivacco Berrone.
Opposite top: skiing up Valfredda. Opposite bottom: departing from Bivacco Berrone.

10 Bivacco Vallonetto (3216m)

Colletto (1630m), Bardonecchia, Italy

2 places (1 hammock, 1 sleeping bag)

No fireplace or permanent stove

Melt snow or collect water on approach

5h00. +1400m & 8.5km from Grange della Valle (Hiking – T4)

From Exilles, park just above Grange della Valle in Vallone di Galambra. Hike W past Rifugio Levi Molinari and continue up to Bivacco Mario Sigot. From there, follow the trail heading W around the plateau to Passo Galambra, then SW towards the rocky towers of Cima del Vallonetto. An exposed and loose path, equipped with cables, leads to Bivacco Vallonetto. This section becomes much more tricky with any snow cover. The hut can also be reached from the SW via the approach routes to Bivacco Berrone.

Tour of Cima del Vallonetto (Hiking – T3)

From Cima del Vallonetto, a scenic hiking loop passes numerous military relics. From Bivacco Vallonetto, head SW past Bivacco Berrone and descend the valley to Ricovero Vin Vert, the ruins of an 1891 barracks built for 310 men, with adjacent stables and a guardhouse. Continue E over Colle dell'Argentera to reach Ricovero Casses Blanches, a two-storey stone barracks from the same era. From there, head NE past another ruined barracks on Monte Chabrière, then N into Vallone di Galambra, returning to Grange della Valle.

Bivacco Vallonetto is a small stone structure built on Cima del Vallonetto, a minor summit on the southern edge of the Ambin massif, overlooking Valle di Susa. From this lofty perch, there is a spectacular view south over the Cottian Alps.

The hut was originally built in 1900 as a lookout post to monitor the Franco-Italian border by the 3rd Alpini, a mountain warfare regiment of the Italian Army. Designed to accommodate three troops, it was used during both the First and Second World Wars. It served as an outpost for the larger nearby barracks, Ricovero Galambra. The original stone plaque still adorns the front of the building, citing the battalion – Battaglione Exilles – along with "Q 3222", the "Quota" altitude, then measured as 3222m. The structure was dedicated to the soldier Vincenzo Bona.

Since 1996, the hut has housed radio equipment and an antenna that relay a communications channel for the rescue services and mountain guides. This system is powered by a solar panel, and a red aircraft warning light flashes on top of the antenna at night.

Above: Valentine approaching Bivacco Vallonetto. Right: Misha beside the hut.
Opposite: Cima del Vallonetto at sunrise, with Pointe Sommeiller behind.

The building was renovated in 2015 to consolidate the structure and make it slightly more comfortable. That same year, it was rededicated to Gianfranco Joannas, mayor of the commune of Exilles, who died in 2008 on the Montagne des Agneaux in the Écrins. More recently, the roof has been replaced, though in winter a few gaps still allow blowing snow to filter in.

Despite the renovations, this remains a decidedly spartan shelter. Inside, aside from the radio equipment, there are just two benches, a table and two small windows. A hammock allows one person to sleep, while another can lie on the benches or directly on the wooden floor. The benefit of this remote setting and basic lodging is solitude – one is unlikely to be fighting the crowds here. Given the hut is uninsulated, it is most suited to warm summer nights. That said, the tranquillity of this area is perhaps best enjoyed in winter or spring, as the distant buzz of motorbikes and off-road vehicles on the Colle Sommeiller road in summer can seem quite incongruous.

We'd turned back just 70 metres from Bivacco Vallonetto in winter 2023, forced to retreat by oncoming weather, but finally reached the hut two years later. After spending the night in Bivacco Berrone, Valentine, Misha and I set off in the morning, zigzagging up the crusty, sun-baked snow on the southwest flank of Cima del Vallonetto before scrambling along the south ridge to reach the hut. Although the bivouac is unsuitable for winter stays, Cima del Vallonetto makes a fine ski objective. The north, west and east faces all offer excellent ski lines.

After a break, we assessed our options. A direct line north seemed possible, but the slope rolled away steeply, and the snowpack felt questionable. Later, we spotted that a colossal avalanche had released from that spot just days earlier – it was far steeper than it had appeared from above. Instead, we continued along the northeast ridge, crossing the plateau to an easy col that gave access to a broad slope of beautiful, low-angled powder. From there, we traversed to Passo di Valfredda, skied another elegant line, and continued our journey towards Colle Sommeiller.

Above: Valentine skiing the northwest flank of Cima del Vallonetto. Top left: Valentine in Bivacco Vallonetto.
Bottom left: Valentine and Misha approaching up the southwest flank of Cima del Vallonetto.

Chasing History across Mont Cenis

Bivacco Hannibal · Bivacco Walter Blais · Bivacco Mario Sigot · Bivacco del Sommeiller

As the sun dipped behind the Dents d'Ambin, a chilly breeze flowed down the valley, bringing relief after the hot, sweaty climb from Bramans. The snow had been sculpted into waves of sastrugi by strong winds, but the mountain ridges were almost stripped bare. We passed uncovered pools of water and a clear stream, still flowing slowly over a bed of gravel even in the depths of winter. It was late February in the French Alps, and Valentine and I had ascended Vallon de Savine, hoping to reach Bivacco Hannibal before dusk.

Until recently, Valentine had spent every winter on race skis, bound to a regimented training schedule for international ski-mountaineering competitions. Since she'd stepped back from that, we could dedicate our winters to exploring remote corners of the Alps. Reaching Col Clapier at the end of the day, we gazed down into Val Susa. I tried to imagine the Carthaginian general Hannibal leading his army and "wagon train" down this precipitous slope.

Historians have searched for centuries to ascertain where Hannibal's army crossed the Alps. For many, it seemed implausible that he crossed any of the higher Alpine passes, let alone with a battalion of elephants. In July 1959, an Oxford engineering student, John Hoyte, set out to test the feasibility of passing Col Clapier, leading an Indian elephant named Jumbo, borrowed from Turin Zoo, across the Alps. They reached Le Planey above Bramans on the way to Col Clapier, but the steep barrier of "La Crosta" proved impassable. Outfitted with leather-soled boots and kneepads, Jumbo was instead guided over the nearby Col du Mont-Cenis – the route that Napoleon favoured for his invasions of Italy in the late 1700s. Jumbo coped fine, returning to Turin in good health despite losing 230kg, almost 10% of her weight. She was revived with cake and Chianti at the arrival party in Italy.

February is typically the coldest part of winter, but in 2023 it was unseasonably warm. The snowpack was

Above: Valentine leaving Bivacco Hannibal at sunrise. Left: Pic Gaspard and La Meije seen at sunrise from Bivacco Berrone.

thin across the Alps, and the ski touring conditions were more like late spring – threading together ragged patches of névé for the sheer pleasure of sliding across the landscape. Restless for adventure, Valentine and I set out for the Haute Maurienne, where the snow had held up better against the warm Saharan winds. We cycled from Bramans to Cernay with skis, boots and overnight gear on our backs, then locked up the bikes and skated along the pistes under the blazing morning sun.

The terrain by "La Crosta" was a chaotic mix of cliff bands and bare slopes where the sun had burnt off the snow. There was too much snow to walk, yet not enough to skin. We moved slowly, struggling over rocks and around trees. After a gruelling battle to gain just 400m elevation, taking off skis a dozen times, we finally reached the easier terrain of Vallon de Savine. Had I been leading an elephant, rather than following Valentine as she broke trail, I might have turned back as well.

In 218 BC, Hannibal's army travelled from Carthage in North Africa, advancing up the east coast of the Iberian Peninsula before battling their way across the Pyrenees and into the Rhone valley by September. By then, they had 25,000 troops and 37 elephants. Hannibal's strategy was audacious – he planned to attack Rome from the north, surprising them by emerging over the seemingly impenetrable barrier of the Alps. Generations of historians have debated their exact route, and Col Clapier fits some of the criteria.

By late October, after numerous skirmishes with Gallic tribes, the army reached the Alps, only to be caught off guard by the brutal conditions. Polybius, a contemporary historian, recorded their struggle: "Winter was now approaching and the early snows were settling on the mountaintops." The descent proved treacherous – hundreds of men slipped to their deaths, and nearly a third of the army perished in the crossing.

Two thousand years later, with the climate warmer and the mountains less glaciated, Col Clapier is more readily accessible. On the Italian side of the pass, Bivacco Hannibal, built in 2014, stands as one of the most beautiful unguarded cabins in the Alps. Constructed from wood with polished aluminium cladding, its irregular hexagonal cross section is reminiscent of quartz, so often found in the Alps. Tapping the snow off our boots, Valentine and I pushed open the door to be met by a rush of warm air. Valentine nodded in silent appreciation as we entered. A solar-powered fan had been heating the hut all day. After a cup of tea, we unpacked our sleeping bags and laid them out in the loft, where a large skylight framed Mont Giusalet, catching the last rays of sunlight.

I sat on the doorstep at sunset, swaddled in a blanket, trying to make sense of the landscape. As the light shifted and my hands froze, I struggled to make a little drawing of the Susa valley and Orsiera mountains to the southeast. The passage of Col Clapier just didn't seem like a particularly logical place for an army to try and cross the Alps.

However, around the hut I noticed relics of wartime fencing, with trails of barbed wire leading up the hillside. Though barely noticeable, several concrete-reinforced slits are burrowed into the cliffs on either side of the col. Hidden behind these is a vast network of Italian tunnels and machine gun installations dating from the 1890s and 1910s which were further refined in 1939. These were built to defend against the possibility of French troops bypassing Col Cenis to attack via Col Clapier. A barracks for 50 soldiers lies just 500m from the hut, out of sight below the brow of the hill. On the French side, similar, though less developed defences are dissimulated into the mountainside. Despite these fortifications, one might still conclude, as Napoleon did, that Col du Mont-Cenis remains a much more logical passage into Italy.

The next morning, we rose early and set off up a steep couloir, then shuffled across hard, crusty snow on the flanks of Mont d'Ambin. We crept up an exposed, icy traverse to gain Col de l'Agnel Supérieur. From the notch, we spotted the tiny red dot of Bivacco Mario Sigot, over 4km away across the valley. Above it, nestled on the skyline, was the concrete shelter at Cima del Vallonetto, built in 1900 as a military observation post by the 3rd Alpini Regiment.

To the west, just hidden from view, was the new bivouac at Col Sommeiller. With all the complex terrain in between, it seemed unlikely that we'd make it that far in one day. We slid down onto the ridge above Col d'Ambin, scrambled along for a few minutes, and arrived at Bivacco Walter Blais, a small red cabin made of wood and fibreglass, built in 1978. I had thought we might spend the night here, but the hut, situated in a notch in the lee of a rocky outcrop, was half buried in snow. We clambered in through the top half of the door, and our hearts sank as we dropped into the frigid interior, heavy with the odour of damp blankets. "Let's hope the hut at Col Sommeiller is open," Valentine remarked as she munched a sandwich.

Continuing west down a ramp, we then climbed up Glacier d'Ambin, grateful for its shade, which offered relief from the baking heat on the sunnier slopes. Other than the thin, white décor, it didn't feel like we were in the middle of winter. From the summit of Pointe Sommeiller, we had a clear view of the large, new bivouac at Col Sommeiller. I had heard that the Mayor of Bardonecchia had symbolically received the keys the previous autumn, but found no more details. Nevertheless, the prospect of a new, warm and well-insulated hut was worth investigating.

Looking down the wild and remote mountain ridge, I tried to imagine the scene from 60 years earlier. Then, the Alps were straining under the golden age of alpine ski installations, and another type of war was being waged across the mountains. In 1958, Edoardo Allemand, an Italian ski instructor from Bardonecchia, had the hare-brained idea to construct a summer skiing station at Col Sommeiller, with slopes and ski lifts extending down Glacier Sommeiller into French territory. To prove that access was possible, he wrestled a four-wheel drive Alfa Romeo "Matta" up the Italian side, envisaging how this could be turned into a road for easy access.

Work began in 1962 with the construction of a 22km dirt road, from Bardonecchia, past the picturesque hamlet of Rochemolles at 1620m to the glacier at 3009m. Ski lifts were installed the following year, along with a hotel-restaurant complex with 80 beds. However, their fortunes soon turned when a massive avalanche destroyed the original buildings in the winter of 1968/69. As the climate warmed over the following decades, snowfall diminished and the glacier shrank. Skiing ceased in the early '80s, and the resort company was liquidated in 1989.

Most of the equipment was abandoned on site. It was only in the early 2000s, in a massive cleanup operation, that the debris of pylons, cables and ruined buildings was cleared off the mountainside. The design of the new Bivacco, with its architecture and burgundy colour, echoes the original hotel. It seemed a curious choice to make this reference – was this history to be celebrated, or lamented?

As Valentine and I surveyed the scene, a warm wind blew over the summit, even at 3332m. I thought of Hannibal and his elephants. Perhaps, if Edoardo Allemand had rattled a motor car up the open mountainside to 3000m, Hannibal's army could have passed over this col. Below Glacier Sommeiller on the French side, a winding ramp of navigable terrain leads up from Vallon d'Ambin. Yet nowadays, in the summer the glacier below the Pointe Sommeiller is sad and grey. Covered in dust and rock, what little remains retreats several metres each year. Even in mid-February, the south side was just a patchwork of rocks and melting snow, desiccated by weeks of warm, anti-cyclonic weather.

To reach Col Sommeiller, Valentine and I descended a steep gully that led onto Glacier Sommeiller. Crossing this deserted, wild mountainside, I struggled to reconcile this with the summer scene half a century ago – men and women wiggling their bums down the slushy summer snow, gliding down the pistes on long wooden skis and beautiful leather boots. I imagined the pop music blaring from the jukebox at the bar on the col. This site was only abandoned as the changing climate rendered it unusable, not because the mentality of the outdoor industry has changed. In Les 2 Alpes, Zermatt and Saas Fee, the resorts continue to groom pistes each summer down glaciers that are slowly dying. Perhaps, in another few decades, even these will be cleared of the concrete bunkers of industrial tourism, and we'll look back at photos from the past with the same surreal sense of disorientation.

Valentine and I skinned up to Col Sommeiller, reaching a bluff overlooking the new bivouac. The building was much bigger than I had anticipated, its wooden structure clad in sheet metal and painted a deep red oxide.

Above: inside Bivacco del Sommeiller, old skis and historical photographs document the ski installations and buildings of the 1960s.

Peering through the windows, we glimpsed beautiful pine panelling and spacious rooms, it looked like a lovely place to spend a quiet night, but the doors were locked.

It was already late in the afternoon, but rather than descend to Refuge d'Ambin and its comfortable winter quarters and wood stove, I thought we might still reach Bivacco Mario Sigot, the red hut we had seen far away from Col de l'Agnel. In the evening light, we threaded our way up sheltered combes, hopping between snow patches on the higher ground. We passed Ricovero Fourneaux, a 12-man lookout barracks built in 1900, and then the massive buildings of Ricovero Galambra, a barracks for 120 men constructed the following year. Giant rolls of barbed wire dotted the landscape like meteors. Leading up to this site were the spartan remains of a cable lift, built in 1940 to transport supplies up from Grange della Valle, 1200m below us at the end of the military road. Against this backdrop, it seemed less incongruous that the ski installations weren't removed for decades; perhaps this is what we call "heritage".

We arrived at Bivacco Sigot, tired and hungry, just minutes before the sun disappeared behind the mountains and the temperature dropped. Valentine set to work, hacking the ice from the door while I lit the stove. Gazing out across the military wasteland, it seemed an inhospitable place to endure year-round. Valentine, an army doctor herself, imagined that it must have felt all the more wretched for soldiers to wage war in such a beautiful landscape, knowing they couldn't truly appreciate it.

Valle di Susa spread out below us, veiled in clouds. Lying below the Franco-Italian frontier, it has a long history of defensive installations. These stretch back to Roman fortifications in Susa, constructed 200 years after Hannibal marched through. From the 7th to 19th centuries, extensive defensive structures were established all over the region, both in the valley and across every high pass that might present a weakness. Hundreds of barracks, observation posts and artillery emplacements were installed in the most inaccessible locations, complete with roads, cisterns and other infrastructure needed to sustain thousands of men stationed year-round in the high mountains. When Italy switched sides during the First World War, most of these fortifications were vacated and disarmed, with the artillery sent to the Austrian front. However, at the outset of the Second World War, they were re-equipped and further developed. Today, these buildings are gradually crumbling into the landscape, along with thousands of kilometres of barbed wire.

Behind the hut, the chain of mountains we had traversed from the Bivacco Hannibal was silhouetted against the horizon. This shelter was the twin of the half-buried Bivacco Walter Blais we had passed earlier in the day, a wooden construction clad in a protective layer of fibreglass. But here, on a more elevated and exposed perch, the wind had scoured away the snow, leaving the interior dry and inviting. Even so, in the bivouac's guestbook, there hadn't been a single entry in five months. Perched alone on a solitary promontory with a sweeping view over the surrounding valleys, it was easy to see why barracks had been built here. Yet I didn't envy the soldiers, stationed through summer and winter across consecutive wars. I pondered how the mountains can mean so many different things to the people who roam among them.

The following morning, we skinned slowly towards Cima del Vallonetto to inspect the tiny concrete lookout, just large enough to sleep three unfortunate soldiers. As the sky turned milky and overcast, we quickened our pace to reach the horizon. At the crest, I could see the edge of the structure, just 70m away along the ridge. With a dark front of stormy weather closing in from the southwest, I tried to reach the building directly, bypassing the scramble along the rocky ridge. Teetering across steep, hard snow felt desperately exposed, the slope dropping away in a series of rock bands into the valley below. I had pushed far enough. Rather than end up like one of Hannibal's men, I delicately retraced my steps. It was time to head home.

As the sky lowered, we hurried back over Col Sommeiller and into France, making careful turns down the glacier in the flat, deceitful light. Here, most traces of the ski industry had been erased. Perhaps this is heritage best forgotten, after all. The legacy of the two world wars was quite enough to think about. We paused briefly for lunch at Refuge d'Ambin, nestled in the base of that austere valley, before battling through squalls of snow back to our bikes at the end of the road, as the mountains disappeared into a veil of cloud.

Left: from Passo dei Forneaux, looking across the ruins of Ricovero Galambra to Bivacco Mario Sigot. Previous spread: Valentine approaching Bivacco Hannibal.

11 Bivacco del Sommeiller (3009m)

Rochemolles (1630m), Bardonecchia, Italy

16 places (no mattresses)

No fireplace or permanent stove

Melt snow or filter water from nearby streams

5h00. +1350m & 14km from Rochemolles (Hiking – T1)

From Rochemolles, hike, ski tour or mountain bike up the 4×4 track, ascending Valle di Rochemolles to Rifugio Scarfiotti, and then continue up Vallone di Fond to Colle Sommeiller. Between July and September, this track becomes a busy toll road, open to motor vehicles for 5 or 6 days a week, all the way to the col. To take advantage of this easy access, or conversely avoid these noisy, crowded days, contact the Bardonecchia tourist information for current information.

Pointe Sommeiller (3332m) Ski Tour (Ski – 3.2)

Pointe Sommeiller offers fine ski touring options, and this loop can be completed in either direction depending on the snow conditions. From Colle Sommeiller, ski SW before ascending E to reach Passo dei Fourneaux. From there, climb the gentle SE flank to Pointe Sommeiller. From the summit, a variety of descents are possible. A fine 250m, 40° couloir descends SW from the summit, rejoining the approach. For extreme skiers, an exposed line (250m, 50-55°, 5.2, E3) descends N from the summit. 400m ENE of the summit, two parallel couloirs (200m, 45-50°, 4.3) split the NW face of Pointe Sommeiller. For a more mellow loop, follow the ENE ridge for 800m to reach the bowl W of Pointe d'Ambin. From there, ski N then W down 35-40° slopes to reach Glacier Sommeiller, and traverse W back to Colle Sommeiller.

Rognosa d'Etiache (3382m) via NE Ridge (180m, PD, III)

Rognosa d'Etiache is one of the most beautiful peaks in the region, formed of a sharp crest of yellow quartzite. From Colle Sommeiller, a short but exposed climb leads to the NE summit, from where you can continue along the ridge to reach the principal SW summit. From the bivouac, ascend the E flank of Rognosa d'Etiache to the NE shoulder at c.3280m. From there, climb 10m, then traverse 30m R (W) to reach the base of a gully. Climb a step into the gully, then ascend a hidden ramp leading L to the crest. Follow the crest R, and climb easily to the NE summit (3373m). From this point, you can descend the NW flank (passing a step of III lower down) towards Rifugio Scarfiotti. To reach the SW summit, follow the crest W over Pt 3376m and descend to the col (c.3320m) between Pt 3376m and the SW summit. From the col, skirt R (NW) around a steep gendarme, then climb a few rock steps (III) to regain the ridge at a notch (c.3340m). Follow the crest to the SW summit. To descend, either reverse the entire climb or descend the NW flank from the NE summit. Alternatively, from the notch (c.3340m) before the steep gendarme, make a 30m rappel (bolts) down the SE side. Make a second 30m rappel heading L, then scramble SW close beneath the SE face until c.3100m. From this point, at the foot of Torre Maria Celeste (which offers a selection of fine 300m rock routes on its SE face), zigzag E down scree to reach the Colle Sommeiller road.

Bivacco del Sommeiller is a sturdy wooden structure clad in painted steel, opened in 2024. It stands on Colle Sommeiller, a broad pass on the Franco-Italian border beside a small lake left by the retreating glacier. The pass is accessible via a 14km dirt road, originally constructed in 1962 to facilitate access to a summer ski resort that once stretched from the col down the glacier into France. In the ensuing decades, as the climate warmed and snowfall decreased, the glacier retreated. Maintaining the lift infrastructure and pistes became increasingly difficult, and the resort finally closed in 1984.

The peak to the southeast, originally known as Rognosa di Galambra, was later renamed Punta Sommeiller / Pointe Sommeiller in honour of Germain Sommeiller, a pioneering engineer and politician of the Kingdom of Sardinia – a precursor to modern Italy. He was responsible for the design and construction of the Fréjus Railway Tunnel, the first major mountain tunnel of its kind. Completed between 1857 and 1871, this 12km tunnel was a remarkable engineering achievement for its time, linking what are now France and Italy.

Built from local wood harvested from the forests of Val Susa, Bivacco del Sommeiller combines traditional materials with a contemporary aesthetic. The structure is made of larch, with pine interior cladding, larch and oak flooring, and poplar furniture. Its design subtly echoes the ski resort buildings of the 1960s, incorporating formal elements and the maroon colour scheme that lend a sense of historical continuity.

The unguarded hut features a spacious dining area with an expansive window overlooking the lake. It sleeps 16 people in two dormitories, though at the time of writing there are no mattresses or blankets. While well insulated, the generous interior volume may be challenging to warm in winter without additional heating.

The refuge stands as a testament to the cultural history of the site. Its corridor displays vintage skis and archival photographs that capture the cooperative spirit needed to run a small community resort in the bygone era of summer skiing. Owned by the town of Bardonecchia, the bivouac remains under development, with a management plan still being finalised. Designed as a single-storey structure, the hut is accessible to visitors with motor disabilities during the summer season.

Top right: Valentine arriving at the bivouac in February 2023, finding it still locked and inaccessible. Bottom right: Valentine and Misha inside Bivacco del Sommeiller in 2025. Note the snow blocking the entrance and window behind them.

SCUOLA ESTIVA SCI
Mt. 3200
SOMMEILLER

When Valentine and I skied past in 2023, the building was nearly finished but was still locked. We returned in March 2025, a week after a major storm cycle had draped the region in fresh snow. After ascending Valfredda and spending the night at Bivacco Berrone, we crossed Cima del Vallonetto, enjoyed a stunning descent into Vallon du Fond, and began skinning up towards Colle Sommeiller.

The hut lay under considerably more snow than two years earlier. The two southwest-facing doors were partially buried, and even after clearing the ice, they appeared locked. This was puzzling, as local guides had assured me the hut was open. The northeast side was entirely buried beneath a deep drift of wind-blown snow. As a last resort, Misha began digging and eventually uncovered a cavity leading to a hidden doorway. Dropping down onto the wooden floor, we were met with the evocative scent of sawn larch and pine. Despite the hut's ideal location for ski touring, the winter entrance design seemed an odd choice. Later, in conversation with the architect, we learned that the building was still in a testing phase and that modifications to the winter access were under consideration.

After lunch, we set off on skis towards Rognosa d'Etiache. Reaching a snowy shoulder beneath the steep summit buttress, Valentine and Misha hesitated at my chosen route, which looked tougher than the grade suggested. While they observed, I climbed ahead to check it out. Moving hastily, I scampered up and traversed right to reach a gully. The terrain wasn't steep, but the unconsolidated snow over slabby rock made for insecure footing. With the north face dropping away below, it immediately felt exposed.

The northeast summit was first climbed by this route in August 1882 by Giuseppe Corrà, guided by Francesco Médail. Of this same passage Corrà wrote: "We encountered a couloir of dizzying steepness, which gave us pause. The fresh snow, unable to adhere to such a steep incline, had settled in the gaps between the rocks. Hardened by frost, it made climbing extremely treacherous."

I carefully climbed an awkward step into the couloir, taking care to leave enough snow to allow me to descend. For such an easy climb, the combination of precarious footing and objectively serious terrain did make it feel rather "treacherous". As I suspected, above was a hidden ramp of snow. I punched deliberate steps, scraping through to the rock beneath, and soon emerged onto the sunlit crest, where I could see Valentine and Misha below, looking up with an air of consternation.

Above: Valentine skiing below Passo di Valfredda.
Left: Misha and Valentine arriving and tunnelling into Bivacco del Sommeiller.

Pointe Sommeiller (3332m)

Deep snow made the ridge laborious, but after climbing a final slope of steep powder, I reached the northeastern point. A jagged ridge led along to the main summit, just a few metres higher, but with the day well advanced, it was time to descend. I retraced my steps carefully and rejoined Valentine and Misha at the shoulder. I had been gone for 40 minutes, and they were getting cold. I hurried to get my skis back on and set off just behind them. Eager to make up for my delay, I set off too fast. Misjudging my first turn in the crusty snow, I caught an edge and face-planted, cutting my forehead on the abrasive surface. Otherwise unhurt, we skied more carefully down the perilous, icy slopes back to the hut.

After gathering our gear, we set off for the long ski back to the car. It was already 5pm, and the cooling temperatures worked in our favour: the snow became safer, even if it was difficult to ski. On the long descent of Valle di Rochemolles, we criss-crossed the valley, navigating every variety of horrendous, crusty snow. Below Rifugio Scarfiotti, we had to pass beneath massive south-facing slopes that were prone to avalanche in the heat of the day. Our timing had been deliberate – later in the afternoon, the snow was refreezing. Even so, we raced beneath the hazard and slid back to the safety of the north side, shuffling along above the reservoir and threading patches of spring snow all the way back to the car.

Above: the north face of Rognosa d'Etiache, showing the NE ridge (left line), NW flank (right line) and the crest route to the SW summit. Top left: Misha arriving at c.3280m at the base of the steep buttress on the NE ridge of Rognosa d'Etiache. Bottom left: Misha and Valentine on the lower NE ridge of Rognosa d'Etiache, with the Pointe Sommeiller behind.
The red line shows the bivouac (circled) and the ski touring loop. The dashed line below the summit marks the N face descent; parallel dashed lines indicate the couloirs on the NW face.

12 Bivacco Mario Sigot (2980m)

Grange della Valle (1830m), Exilles, Italy

8 places (8 mattresses, 10 blankets)

No fireplace or permanent stove

Melt snow, or filter water from Lago Galambra.
Both sources may dry up completely during summer.

4h00. +1170m & 5.5km from Grange della Valle (Hiking – T2)

From Exilles, park just above Grange della Valle in Vallone Galambra. Hike W past Rifugio Levi Molinari, then continue uphill past Lago delle Monache and the ruins of the WWII cable car station to reach the bivouac.

Tour of Pointe Sommeiller (Hiking – T3)

This hike can be linked with the Tour of Mont d'Ambin to create a 3-4 day circuit, using Bivacco Mario Sigot, Walter Blais, Hannibal or Sommeiller. See the Bivacco Hannibal chapter for further details. From Bivacco Mario Sigot, head NW to Passo Settentrionale dei Fourneaux. Descend W, then follow the track NE to Bivacco del Sommeiller at Colle Sommeiller. From there, descend NE then N past Glacier Sommeiller. Pick up a better path and follow this down to the trail junction at 2350m in Vallon d'Ambin. Turn SSE and ascend to Lac d'Ambin. From the lake, climb SW via a broad gully (often snow-filled in early summer, 35-40°), then head ESE to Bivacco Walter Blais at Col d'Ambin. From there, descend SSE on a steep path. At 2537m, either traverse SW to return to Bivacco Mario Sigot, or continue descending to Grange della Valle. In stable conditions, this circuit also makes an excellent ski tour.

Pointe Sommeiller (3332m) via South Ridge (Hiking – T3)

From the bivouac, a marked trail winds up the desolate, rocky flanks of Pointe Sommeiller. Head NW past the large Galambra barracks, then W towards the lookout post at Passo Meridionale dei Fourneaux. Continue NE, then N, ascending a long gravel ridge to reach the summit (1h30 from Bivacco Mario Sigot). Descend by the same path. This route also makes a simple ski tour, though the snow tends to melt fast in spring on the S-facing slopes.

Bivacco Mario Sigot occupies a magnificent position, perched on the rock lip at the edge of the glacially carved hollow below Pointe Sommeiller. It offers sweeping views of the Mont-Cenis and Orsiera ranges, extending far across the plains to Torino. Sunrise from the hut can be spectacular, and on a clear night, the lights of the towns far below feel satisfyingly remote.

The bivouac was built in 1997 and named in honour of Mario Sigot, a mountaineering instructor from the CAI section of Susa, who died in the spring of 1994 on Grand Hoche, a mountain on the French border at the head of Valle di Susa. Like Bivacco Walter Blais, the hut is constructed from a metal frame with a wooden lining and fibreglass cladding.

Above: Valentine skiing towards Passo dei Fourneaux.
Right: Valentine enjoying a cup of tea at dusk outside Bivacco Mario Sigot.

CLUB ALPINO ITALIANO
SEZIONE DI SUSA
BIVACCO
MARIO SIGOT

Scattered across the surrounding mountains are countless barracks, gun emplacements, lookout posts and other relics of the region's defence network. These were built between 1890 and 1940 in anticipation of a French invasion that never came. Today, they stand as stark reminders of a very different era and of how fortunate we are to have known 80 years of peace.

Situated on the edge of the plateau between Pointe de Sommeiller and Cima del Vallonetto, Bivacco Mario Sigot makes an excellent base to explore both peaks. Each offers a rewarding ascent and even grander vistas than those from the bivouac. From Pointe Sommeiller, the panorama stretches from Monviso, 55km to the south, across the Écrins to the southwest and the Vanoise to the northwest. On a clear day, even Mont Blanc is visible, 80km to the north.

Until just a few decades ago, Lago Galambra, behind the bivouac, was considerably larger. It formed in a depression at the snout of the Galambra glacier (Ghiacciaio del Galambra), which has since vanished. In the late 19th and early 20th centuries, this glacier played a key role in the regional ice trade.

As living standards rose, hotels and tourist establishments were increasingly expected to provide guests luxury goods such as beer, soda and frozen desserts that needed to be kept cold, using ice. As ice melted quickly down in the heat of Turin, an efficient chain of transport was imperative. In Piemonte, the only glacier that was both easily accessible and close to a railway station was at Passo Galambra, 2000m above the village of Salbertrand.

Workers cut the ice into 300kg blocks from the steep eastern front of the glacier. A team would then leave the valley at midnight, ascend to the glacier and manoeuvre a block onto a sturdy 30kg sled. From there, they dragged it south to Pasòu du Gla (the ice passage), then descended the gorge of the Rio Geronda to reach the mule track at the village of Frenèe, and finally continued down to meet the train at Salbertrand. Between extraction at the glacier and arrival at the train, around 15% of the ice typically melted. The remaining blocks were weighed, and then loaded into thermally insulated wagons for transport. These operations continued throughout the summer. The effort and risk involved in controlling a 300kg sledge careering down the rough, dry mountainside is rather hard to imagine nowadays.

Top left: Valentine clearing ice from the door of Bivacco Mario Sigot. Top right: preparing dinner.
Opposite: Valentine leaving Bivacco Mario Sigot in the morning, with a cloud inversion blanketing the valley below.

13 Bivacco Walter Blais (2900m)

Grange della Valle (1830m), Exilles, Italy
or Pont d'Ambin (1895m), Bramans, France

8 places (8 mattresses, 20 blankets)

No fireplace or permanent stove

Melt snow, or filter water from Lac d'Ambin

3h30. +1070m & 4km from Grange della Valle (Hiking – T3)

From Exilles (Italy), park just above Grange della Valle in Vallone Galambra. Hike NW up an ever steepening path, equipped with fixed ropes, to reach the bivouac at Col d'Ambin.

3h30. +1040m & 8.5km from Pont d'Ambin (Hiking – T3)

From Bramans (France), park at Plan de la Vie (1850m) or Pont d'Ambin. Hike up Vallon d'Ambin, passing Refuge d'Ambin and Lac d'Ambin. Above the lake, head SW up a broad gully (often snow in early summer, 35-40°), then follow a shelf ESE to reach the bivouac at Col d'Ambin.

Pointe Ferrand (3365 m) via West Ridge (F, I+)

This route provides an easy ridge scramble and snow climb over the remnants of Glacier Ferrand. Note that on most Italian maps, Pointe Ferrand and Pointe Niblè are reversed. This ascent refers to the Pointe Ferrand as marked on French IGN maps, the SW of these two summits. From the bivouac, scramble E up the ridge to c.3050m. Continue NE up the snow or scree along the edge of Glacier Ferrand, keeping L of the cliff bands on the crest. Head back R (E) to regain the W ridge via a scree or snow slope, and follow the SW ridge to the summit (2hrs from Bivacco Walter Blais). Descend via the same route. Alternatively, to make a loop back to your departure point, descend the summit ridge and head N across Glacier Ferrand to Col de l'Agnel Supérieur. Continue ENE, down towards Rifugio Vaccarone. From there, either head SW via Passo Clopacà to reach Grange della Valle (Italy), or N via Col Clapier, Bivacco Hannibal and Vallon de Savine to reach Pont d'Ambin (France).

Bivacco Walter Blais is perched at Col d'Ambin, a notch on the Italian–French border in the very heart of the Massif du Mont-Cenis. While this region is most commonly visited via the road over Col du Mont-Cenis, which provides easy access to the neighbouring valleys, the head of Vallon d'Ambin, just one valley further west, remains largely deserted. If the accessibility of Vallon de Savine and Bivacco Hannibal in summer makes it unappealing, then Walter Blais provides an excellent alternative for quieter trekking and wild alpine scrambles.

This bivouac was erected in memory of Walter Blais, a gifted young mountaineer from Susa. He died in May 1978, at the age of 20, while marking out the course for the "Trofeo Penne-Mozze" ski mountaineering race, not far from the bivouac site.

Built from a metal frame with wooden lining and fibreglass cladding, the structure offers better insulation than the older tin bivouacs. Inside, there are bunks and mattresses for eight people, along with a small table and benches for cooking.

In summer, the hut feels comfortable and spacious, especially considering its age. In winter, however, it can become entirely buried in snow. Then, like a snow cave, it becomes cold and slightly damp. The outward-opening door also poses one significant hazard in winter: you can be blocked inside by snowfall. Fortunately, a small trapdoor has been installed to allow excavation from within if necessary. If planning a visit in winter or spring, it's worth bringing a probe, a good shovel and possibly a sleeping bag. While the area is well suited to ski mountaineering, the winter room of Refuge d'Ambin might be a more comfortable base. Autumn is perhaps the most attractive time to visit, when steep snow patches no longer hinder access.

The geology of the Massif du Mont-Cenis is highly varied, though mica schist and quartzite dominate, resting on a base of gneiss. Much of the rock is poor for climbing, but there are impressive exceptions, such as the magnificent Grand Bec d'Étache, a steep face of quartzite in the neighbouring valley to the west. Closer by, Le Grand Cordonnier offers a fine scramble at PD+.

Despite the steep and rugged terrain, historical records show that for centuries Col d'Ambin was a common route for trading livestock. By the late 19th century it had become a favoured crossing point for smugglers moving salt and cloth across the border. Some Italian reports even speak of entire flocks of sheep being stolen and driven over the pass into France. Today, you are more likely to encounter herds of ibex or chamois in this secluded region than smugglers or sheep rustlers.

Above and left: a shorter stop than planned as we found the hut half buried and rather damp!

14 Bivacco Hannibal (2477m)

Col du Petit Mont-Cenis (2183m), Val-Cenis, France or Pietra Porchera (1154m), Valle Clarea, Giaglione, Italy

8 places (8 mattresses, 18 blankets)

No fireplace or permanent stove

Spring 300m ESE in summer. Stream 250m SW in winter

USB ports, lighting and a heating fan are powered by solar panels

2h30. +400m & 7.5km from Col du Petit Mont-Cenis (Hiking – T1)

From Col du Mont-Cenis (France), follow the road to Col du Petit Mont-Cenis (generally open from June to October). Hike SE to Lac de Savine, then continue to Bivacco Hannibal at Col Clapier. Alternatively, access is possible from Bramans, parking at Le Planey or Plan de la Vie (1850m).

4h00. +1350m & 7.5km from Pietra Porchera (Hiking – T3)

From Giaglione (Italy), drive up Valle Clarea and park at Pietra Porchera (1154m). Follow the mule track NW to c.1360m, then continue steeply up to Col Clapier and Bivacco Hannibal.

Tour of Mont d'Ambin (Hiking – T3)

From the bivouac, follow the path S across a valley and up to Rifugio Vaccarone. Continue WNW over Col de l'Agnel Inférieur for a shorter route, or via Col de l'Agnel Supérieur, via Bivacco Walter Blais for a longer loop. Descend NW into Vallon d'Ambin and then complete the loop E back to your starting point.

Mont d'Ambin (3378m) Ski Tour (Ski – 3.2)

The region offers excellent ski touring, and winter is perhaps the best time to visit the bivouac in peace. Access is possible via the lifts at Col du Mont-Cenis or the ski de fond tracks above Bramans. In winter, the Vallon d'Ambin road is closed just above Bramans, but a shuttle bus operates to Le Cernay (1600m), where you can follow tracks up the valley. From there, climb "la Crosta" towards Col du Petit Mont-Cenis – tricky when snow-covered, though it often clears early in spring. Head SE into Vallon de Savine to reach Bivacco Hannibal. From there, either traverse and climb S near the summer path (45°, fixed ropes) to Rifugio Vaccarone, or boot up the gully WSW of Lago Clapier (35–40°). Ski WSW to reach Col de l'Agnel Inférieur (35°). From there you can make the optional ascent of Mont d'Ambin (3378m) via the SE ridge or SW flank. The tour can be extended in many ways (see options in Bivacco Mario Sigot chapter), or to complete the loop, descend W into Vallon d'Ambin, past Refuge d'Ambin, which has an excellent winter room with a wood stove, and return to Le Planey.

Mont Giusalet (3312m) via SSE Ridge (PD, III)

This route provides an easy ridge scramble with a short passage of climbing on good rock. From the bivouac, ascend E to Col de la Vecchia (2971m), then follow the ridge NNW. The crest soon narrows, offering varied and fun climbing, before joining the normal path from Rifugio Avanzà. Follow this to the summit (3hrs from Bivacco Hannibal). From Mont Giusalet, either reverse the route or descend the path SE then E towards the Col Cenis road and walk or hitch back to your starting point. Alternatively, descend NE from the summit via the scree amphitheatre to reach Bivacco Piero Vacca, then hike back to your starting point.

Col du Mont-Cenis has been a major Alpine crossing for centuries. Until the 18th century, carriages had to be dismantled at the base of the pass and carried over by mules. The first modern road over the col that was suitable for carriages was constructed by Napoléon in 1802. This was his preferred route to the Italian plains, having used the pass during several of his military campaigns in the 1790s. Today, the road is maintained open from May until November.

From 1860 to 1947, the Col marked the frontier between Italy and France, and the surrounding area is still dotted with historic military fortifications. Since the 1947 Paris Peace Treaties, the watershed has been entirely in French territory. Over the past 50 years, particularly with the opening of the Fréjus road tunnel, the Haute Maurienne and Mont-Cenis area have shifted from a critical transport axis to a tourism destination.

With its grand turquoise lake and gentle alpine meadows, the massif is ideal for summer hiking and ski touring in winter. From June until October, the small dead-end road to the higher Col du Petit Mont-Cenis is usually open. The "GR de Pays du Tour de Haute-Maurienne Vanoise" also passes through here, a 154km hiking trail that loops around the region.

Above: the view to the north. Right: Valentine outside Bivacco Hannibal at dusk.

Col du Petit Mont-Cenis also provides access to Vallon de Savine, where Bivacco Hannibal, one of the most beautiful unguarded cabins in the Alps, stands watch over the passage into Italy. Erected in 2014 as a cross-border collaboration between the municipalities of Bramans (France) and Giaglione (Italy), the hut commemorates Hannibal's legendary Alpine crossing.

Due to its accessibility from Col du Petit Mont-Cenis, Bivacco Hannibal can be busy in summer, especially with day hikers. That said, you're unlikely to find Hannibal's entire army camped there. Once the roads close with the first snowfalls of winter, the journey becomes far more demanding and solitude much more likely.

The entrance is positioned on the south side, yet curiously, the only downstairs window, a large panoramic opening, faces north and overlooks the flat expanse of Vallon de Savine. Its orientation seems to deliberately forgo the sunlight and dramatic views over the Susa valley and Orsiera mountains to the southeast.

Clad in polished aluminium, the bivouac's striking form, an irregular hexagonal prism, echoes the geometry of a quartz crystal. The building is constructed from laminated timber, with a large main dining room downstairs and a ladder leading to a compact sleeping loft above. This features a skylight and a gently sloping roof that helps prevent snow accumulation.

The interior feels spacious and comfortable, and the generous tables could host a dinner party of twenty. For overnight accommodation, there are eight mattresses: six in the loft and two fold-down bunks downstairs. On sunny days, solar heating pumps warm air into the hut. Combined with excellent insulation, this makes it surprisingly cosy in winter, despite its altitude and exposed location.

Above: Valentine outside Bivacco Hannibal at sunrise.
Top left: Valentine cycling to Le Cernay. Top right: looking out the north-facing window. Centre left: in the loft.
Centre right: Valentine skiing up La Crosta. Bottom left: skiing up Vallon de Savine. Bottom right: settling in at Bivacco Hannibal.

Landscape of Conflict

Refuge du Camp des Rochilles

Straddling the border between Savoie and the Hautes-Alpes, the Massif des Cerces stretches south and east, flanked by the bustling Guisane valley and remote Vallée de la Clarée. This wild region is marked by striking contrasts. Over its foundations of schist and sandstone protrude limestone and quartzite ridges, sculpted into fantastic shapes. The cultural landscape is equally diverse, balancing a fragile agrarian economy, past and present military installations, and an ever-growing tide of mountain-loving tourists, hungry to devour the rough tracks and startling views.

Situated at the southern edge of these mountain ranges, below Col de Montgenèvre, a historic and strategic crossing point of the Alps, Briançon has been fortified since the Middle Ages. After battles erupted in Europe at the turn of the 18th century, the French engineer Vauban commissioned a vast system of defensive installations along the French border, transforming Briançon into a key military frontier town.

Tensions with Italy flared once more when it joined the Triple Alliance at the end of the 19th century, prompting a significant reinforcement and expansion of border fortifications to accommodate advancements in artillery technology. The garrison at Briançon grew substantially, from around 1000 troops in 1850 to nearly 4000 in 1914. During this period, a significant military base was established at Camp des Rochilles on the boundary between the Massif des Cerces and the Massif du Thabor, with barracks and supporting infrastructure constructed between 1902 and 1907. The nearby pass, Seuil des Rochilles, was identified as a strategic vulnerability in the event of an Italian attack launched from Bardonecchia.

After a period of relative calm following the First World War, the high-altitude installations along the

Above: Valentine riding past Lac de la Ponsonnière. Left: riding over Col des Cerces. Previous spread: riding at sunset beneath the Arêtes de la Bruyère.

frontier were reoccupied in the late 1920s. During the 1930s, a renewed wave of defensive construction was undertaken as part of the "Maginot Line", which spanned the entire eastern border of France. This time, the defences were designed as multiple staggered strongholds of resistance. The rugged crest of the Massif des Cerces and Thabor still bears the remnants of this era, with 149 bunkers, observation posts and gun installations scattered across the landscape.

When passing through these mountains, it's easy to remain oblivious to the traces of this sombre military history. However, at almost any location in the region, half-hidden square-cut concrete eyes are looking down on you. Many of these bunkers remain accessible, and it's both fascinating and eerie to peer through a gunner's stepped embrasure. This shift in perspective offers a window into that distant era marked by fear and aggression. It can be a humbling experience, offering a gentle reminder to cherish the relative peace, cooperation and political stability that Western Europe has enjoyed for several generations.

In early November, just a couple of weeks before the scheduled "Cerces" military exercises of 2024, Valentine and I set off in the late morning. From the end of Vallée de la Clarée, we rode up the road to Laval and the bumpy track beyond. Once above the treeline, the valley offered sweeping views to the Pointe des Cerces and its array of satellites and gendarmes. While a few of these quartzite and limestone formations have fine climbing routes, most are tottering towers of choss.

We pedalled slowly up the rugged track, labouring past scattered groups of walkers. Beyond the Refuge de Drayères, the trail steepened, forcing us to push and haul our bikes over rocky steps.

REFUGE DES ROCHILLES

The occasional hikers looked at us askance, making me wonder why we hadn't chosen to trek through the region like everyone else. Approaching the Seuil des Rochilles, we passed under the looming geometric form of a World War II bunker, one of the few above-ground elements of the largely subterranean network of the "petit ouvrage des Rochilles".

The door was ajar and I ventured in some way before getting spooked by the echoes rattling down the abandoned tunnels. The galleries still hold rusting bunk beds, as if waiting for a ghost army that never arrived. Begun in 1931, the site consists of a network of tunnels connecting three gun emplacements, built to defend the easterly, Italian-facing flank of the Rochilles pass. By the late 1930s, the complex was equipped to house up to 40 soldiers, but aside from a minor Italian incursion further down the valley at Plampinet, these remote outposts never saw active combat.

—

Today, much of this mountainous region is classed as a military shooting range. Based out of the historical Camp des Rochilles, the Grand Champ de Tir des Alpes (GCTA), serves an extensive zone for artillery and infantry live-firing and training exercises. Unique in Europe, it is the only such military training area on the continent not solely owned and exclusively used by the army. In recent years, the training has also gained an "inter-allied dimension" as the French cavalry, artillery and infantry collaborate with their Italian counterparts.

For the young recruits, the live-firing exercise is an invaluable experience, operating in real conditions over demanding terrain in a breathtaking landscape. "2,200 meters above sea level in winter temperatures, Sergeant Aubin prepares the cannon of his AMX-10RC [tank]. It is five o'clock in the morning when the squadron leaves their camp. After a short and freezing night, the priority is to check the machines and equipment to begin the mission. All the vehicles must be camouflaged to blend into the white scenery of the Alps." Through day and night, the crash of explosions resounds across the valley.

Colonel Flore, commander of the 4th RCh, highlights the importance of this exercise: "'Cerces' prepares us for possible high-intensity interventions in conflict zones in Eastern Europe, which are full of mountainous areas (Caucasus, Lebanon, Balkans), or in Arctic areas that are increasingly contested between powers."

Access is prohibited to the entire area during military operations and there are troops posted at all access points to ensure people don't unwittingly head into an active combat training zone. The 155mm CAESAR cannons can fire up to 40km with such precision that 99% of shells land within the area of a football pitch. Only a few limited zones in the Champ de Tir receive these projectiles.

In the summer after each exercise, a massive cleanup operation sweeps the area to remove debris, usually amounting to 2.5 tonnes. For the rest of the year, public access resumes. Although the military uses the region only once per year, its unique classification as a "firing range" has paradoxically preserved it from the development of ski infrastructure. The Cerces massif is now one of the wildest mountain regions of France.

—

By the jumble of blocks at the Seuil des Rochilles, an ancient passage between Maurienne and Durance valleys, a carved stone marker denotes the former boundary between the Kingdom of Piedmont (signified by a cross) and the Kingdom of France (signified by the Fleur-de-Lys). As we dragged our bikes through the boulder field, Valentine began to recognise the terrain. Having served as a military doctor in the mountain artillery regiment, she had participated in numerous exercises from the Camp des Rochilles, and the landscape stirred vivid memories. "Laurent used to love it up here. With his troops, he could do as he pleased. In this remote spot, he was his own boss." Valentine's husband, Laurent Fabre, was a mountain guide and military instructor stationed at Briançon and later in Chamonix. He died in a mountaineering accident in the Mont Blanc massif in 2012 while guiding army cadets.

Emerging from the chilling shade, we were met by a stunning view over Lac Rond. Recently frozen, the luminous surface cracked and vibrated in the November sun.

Top left: horse riders near Refuge des Drayères. Top right: the bunker by Seuil des Rochilles. Centre left: Valentine passing Lac Rond. Centre right: departing from Refuge du Camp des Rochilles. Bottom left and right: the bunkers at Col de Buffère.

Scattered remnants of rusting ironwork and crumbling concrete stood as incongruous reminders of the mountain range's split personality. Back on our bikes, we followed a fine trail leading to Lac du Grand Ban, over the Col des Rochilles and finally down to the military base.

The Camp des Rochilles is nestled at the base of Pic de l'Aigle, one of the region's most spectacular limestone cliffs, its precipitous west face laced with demanding climbing routes. Austere and functional, the main stone buildings are arranged in three grand rows, with an additional long structure offset to the north. One end of this outlying building, originally designed as stables, now houses a room left open by the military as a mountain shelter for public use. Valentine explained that the sturdy red posts bordering the gravel track were installed to guide the tracked vehicles when snowdrifts obscured the path.

We paused for lunch, resting on a mossy bank overlooking the Camp, and spotted a tiny ermine frolicking in the grass a hundred metres away, already clad in his white winter coat. Along with the chamois, ibex and birds we had seen, it was a sign that the ecosystem was less disrupted than one might expect in a military training area.

Leaving the Camp, we ascended back towards the Col des Cerces, riding delicately up a steep north-facing path that merged with a frozen stream. Emerging back into the sunlight, the view spread to the south revealing a landscape of rifts, plateaus and gorgeous lakes scattered across the intricate boundary between quartzite, limestone and schist. The day was wearing on, the light growing softer – a boon for photography but also a clear sign that we were moving much slower than anticipated.

From Col de la Ponsonnière, a scenic path led us across the sun-baked mountain flank, past the eponymous lake and down towards the shaded blade of the Arêtes de la Bruyère. Just weeks earlier, these alpine pastures would have been a hive of activity, with herds of sheep roaming across the mountain under the watchful protection of guard dogs, a response to the predation by an ever-growing wolf population. The fine-looking shepherd's hut stood shuttered and locked, the flocks having already descended to the valley for the final grazing of the season.

Late autumn offers a final, fleeting opportunity to venture into the Alps on foot or bike, as the mountains grow quiet before the onset of winter. By early November, the autumn colours were just beyond their peak, and a summer of herding had left the grasslands neatly trimmed, like a golf course. It was a rare chance to experience the region undisturbed, without being harried by aggressive sheepdogs or bothered by the abundance of our fellow bipeds.

The last rays of sunlight flamed on the hillside as we passed the freshly built Refuge du Clot des Vaches. Constructed with local timber, the sweet scent of freshly sawn larch lingered in the evening air. We were hours later than expected, so we rolled down into the valley as dusk fell, navigating trails by headlamp back through the woods to Briançon for the night. The following day Valentine had to work, so I set off alone to complete the loop. Climbing back up the valley, I crossed the Col de Buffère. Rolling down the north side, I stumbled upon the domed heads of concrete bunkers overlooking Vallée de la Clarée. Started in 1937, these were unfinished by the time of the ceasefire of June 1940.

Curiosity drew me in, and with a headtorch I clambered down the wet concrete steps, my footfalls reverberating through unseen corridors. Damp tunnels stretched deep underground, leading to firing chambers as well as an infirmary and sleeping quarters. It must have been a bleak existence, especially in winter, watching and waiting. Shaking off the grim reverie, I scampered out into the sunshine.

Back on my bike, the narrow, technical trail drew all my attention, leading me down into the Buffère valley. I passed a refuge and a cluster of deserted summer dwellings set in a peaceful alpine meadow, pausing only to poke my head into the well-kept Chapelle Saint Ignace. Descending into the woodland, I finally reached the car and closed the loop we started the day before.

Left: Valentine passing Lac du Grand Ban. Above: descending towards Lac des Cerces.

15 Refuge du Camp des Rochilles (2412m)

Plan Lachat (1960m), Valloire, Maurienne, France

6 places (5 mattresses, no blankets)

No fireplace or permanent stove

Intermittent spring 80m E

1h30. +450m & 3km from Plan Lachat (Hiking – T1)

From Plan Lachat, follow the track SE to Les Mottets. Continue ENE up the track to Col de la Paré, and the Camp des Rochilles just beyond.

3h00. +500m & 8km from Laval (Hiking – T2)

From Laval, at the end of Vallée de la Clarée (parking is restricted in summer), follow the track NW to Refuge de Drayères. From here, take the path NW then W over Seuil des Rochilles to Camp des Rochilles.

Tour of the Massif des Cerces – 40km, +1800m (Hiking – T2)

The Cerces offer a spectacular hiking or technical mountain biking loop. Adaptable to a 2 or 3-day journey, the route passes several of the region's guarded refuges – Drayères, Clôt des Vaches, Buffère, or the unguarded Refuge du Camp des Rochilles. The tour can be accessed from Plan Lachat, Vallée de la Clarée or Pont de l'Alp. Starting from Vallée de la Clarée, follow the track NW to Refuge de Drayères. From there, take the rocky path NW then W over Seuil des Rochilles (sections require pushing your bike). Climb S over Col des Cerces and down to Lac des Cerces. Ascend S over Col de la Ponsonnière then SE past the Lac de la Ponsonnière. Under the E flank of Bruyère, the trail descends to l'Alpe du Lauzet. From here, head SSW up the Chemin du Roy, traversing high across the hillside to reach Col de Buffère. Descend NNE to Refuge de Buffère, then follow the path along the W side of Ruisseau de Gardiole to return to the Vallée de la Clarée.

Pointe des Cerces (3098m) via NW Flank (Hiking – T3)

This easy alpine summit is one of the finest hikes in the Hautes-Alpes, offering a panoramic view across the region. From Camp des Rochilles, follow the path E then S to Col des Cerces. From the col, ascend SE across the flank of Pointe des Blanchets to reach the N cirque of the Pointe des Cerces. Ascend this SE to reach the summit (2h30 from Camp des Rochilles). In anything other than dry, mid-summer conditions, the cirque may retain névé, making an axe and crampons necessary. This outing also makes a great ski tour.

Camp des Rochilles is a military base constructed between 1902 and 1907 to defend the Maurienne valley from a possible Italian advance via the Vallée de la Clarée and over the historic pass of Seuil des Rochilles. Nestled in a natural bowl below the defensive threshold of Col des Rochilles, the camp occupies a fine strategic position, easily accessed by a road from below and protected from avalanches above.

The refuge, located at the east end of the long building at the northern end of the camp, is a tall, rectangular room equipped with a window, tables, chairs and three double bunks. The thick stone walls, while robust, are uninsulated, making the space damp and chilly in winter. However, in summer, it provides a welcome and practical shelter.

Before the First World War, there were already seven substantial stone buildings on the site, three wooden buildings and 15 semi-permanent tents. At that time, the stone buildings were constructed with flat roofs. Six of these original buildings remain standing, including the one now housing the public refuge. They have all been upgraded with steep metal roofs to better shed the heavy accumulations of winter snow. The largest buildings were originally designed for accommodation, with additional areas for catering, logistics and repairs. The refuge building, once a stable, was partially converted into washrooms, though these areas are now mainly used for storage.

Though the public shelter is somewhat stark and not the most picturesque bivouac in the Alps, the Camp des Rochilles and the history of the region are fascinating. For all but the hardiest visitors, a day trip to Rochilles is recommended, with a more comfortable overnight stay at Refuge des Drayères or Clôt des Vaches. Alternatively, bivouac camping is tolerated in the Massif des Cerces from 7pm to 9am, offering perhaps the best way to immerse yourself in this mountain landscape.

Above: Valentine in the Refuge du Camp des Rochilles. Top right: cycling above the Camp des Rochilles. The refuge is at the right side of the building in the background. Right: view of the Pic de l'Aigle towering over the Camp des Rochilles. Following spread: Valentine skiing towards Refuge de Chalance.

DAUPHINÉ ALPS

Saint-Jean-
de-Maurienne
19/ Habert des Jarlons
18 / Habert de la Pierre
Grenoble
17 / Habert des Sabottes
Le Bourg-
d'Oisans
La Grave
DAUPHINÉ
ALPS
16 / Refuge de Chalance
La Chapelle-en-
Valgaudémar
Voiron
Moirans
Crolles
Theys
Villard-Bonnot
Domène
Saint-Égrève
Lans-en-
Vercors
Villard-de-Lans
Vif
La Mure
Corps

Val Cenis
Modane
23 / Rifugio Santa Maria
11 / Bivacco Colle del Sommeiller
14 / Bivacco Hannibal
13 / Bivacco Walter Blais
Susa
12 / Bivacco Mario Sigot
10 / Bivacco Vallonetto
15 / Refuge des Rochilles
Bardonecchia
09 / Bivacco Berrone
07 / Bivacco Lago Laus
08 / Bivacco Robinet
Sauze d'Oulx
Cesana Torinese
Sestriere
Briançon
06 / Bivacco Tornior
COTTIAN ALPS
L'Argentière-la-Bessée
05 / Bivacco di Punta Venezia
Guillestre

Highlands of Valgaudemar

Refuge de Chalance

In mid-February, any journey into the heart of the Écrins must be earned. There are no lifts here, and under a mantle of winter snow the mountains feel larger, wilder and more remote. Valentine and I headed to Valgaudemar to make the most of a brief window of stable weather, and were joined by Silvia and Stefano, mountain guide friends from the Dolomites, who are always keen for a rugged adventure. Given the remote location of the Refuge de Chalance, apart from the occasional chamois, I had little doubt that we would be alone. From La Chapelle-en-Valgaudemar, we passed the "road closed" barrier at Le Bourg in trainers, skis strapped to our packs, and quickly discovered why the route was off-limits to cars in winter. The hillside above was shedding rocks and avalanche debris, leaving gaping craters in the tarmac.

A little further along, the road had been washed away by a landslide, and the remaining tarmac overhung the void. We crept past, furtively gazing upwards in case any rogue rocks tumbled down. From Rif du Sap we trotted up a steep, narrow path, rapidly gaining height. By early afternoon, the air temperature was soaring, and we were sweating profusely in T-shirts. A couple of hours later, we reached the snow line near Cabane de la Muande at over 2000m and finally clipped into our skis.

An old skin track, with a trench of melted footprints on top, suggested a couple of teams had passed through recently. To avoid a steep slope of rotten snow, we kicked steps up a short gully and emerged just 100m from the hut. To our surprise, we spotted four guys outside. They looked as stunned as we were and quickly made room for us.

Above: Valentine by the Refuge de Chalance at dusk. Right: Valentine skiing the north side of Brèche de la Muande.

We learned they had driven from Paris and hiked up the previous day and the unseasonal heat had already melted their tracks. They had come seeking a long weekend adventure among friends but hadn't realised quite how different the high mountains become in winter, even when sticking to marked trails.

Setting off from Le Bourg at 9am, they had taken 12 hours to reach the hut, arriving after dark. Their slow progress was partly due to their footwear: fur-lined rubber boots and running spikes instead of crampons or snowshoes. Renaud, the most experienced in their group, had broken trail the entire way through deep, wet snow, often sinking to his waist. Nathan, exhausted and cramping, had to abandon his rucksack, which the others dragged up on a small sledge they'd brought for the descent. We listened to their story with our mouths agape, impressed and horrified in equal measure. Several of them were still so fatigued that they had decided to stay a second night. Fortunately, their colossal backpacks were stocked with more than enough supplies for their unplanned extension.

In the afternoon, Valentine and I skied up to the Col de Chalance to view the upper section of the northwest couloir of Les Rouies, which we hoped to climb the following day. I lingered at the col for half an hour, hoping the sun might break through the veil of high cloud. Back at the hut, we found everyone settled in with hot drinks and card games. The gas hobs roared like a blast furnace, rapidly melting snow for water, while also warming the cabin. With eight people crammed inside and damp gear strewn everywhere to dry, the hut felt small. We passed the evening in the warm glow of mountain camaraderie before retiring early.

The Parisians, facing a challenging descent in floppy boots on refrozen snow, crept out into the darkness at 6.30am. A couple of hours later, we set off, winding our way up to the ridgeline at Brèche de la Muande to locate the hidden gully giving us access to the north-facing glacier.

In February 1918, Émile Henriot and Camille Simond completed the first ski traverse in the area. They crossed the Brèche de l'Olan and the Col de la Muande, adjacent to the notch where we now stood. Four years into the First World War, with no end in sight, the pair were on leave from the army. The richness of Henriot's writing and the poignancy of their experience in such a grim context served as an evocative reminder for me to pay attention to the details and the immense beauty we sometimes take for granted. They set out from Chapelle en Valjouffrey, where the native shepherds they met knew nothing of the next valley.

Left: Stefano, Silvia and Valentine ascending the northwest couloir of Les Rouies. Above: Silvia and Stefano preparing to leave Refuge de Chalance.

Henriot wrote: "On the other side, where we were going was unknown to everyone there. We had the impression of being at the edge of a mysterious land, finishing at the austere walls of l'Olan ... As we climbed, the mist gently lowered and, for the first time, suddenly, l'Olan appeared, still cloaked in a vaporous scarf. In as little time as it took the wind to blow away a cloud, our two souls, dark and indecisive like the mist surrounding them, overflowed with life, joy and hope. Such glorious changes turn the whole being inside out. These are moments of sublime, regenerative ecstasy that can only be experienced in the high mountains. There is no more striking reminder of humility than these contrasts that are suddenly discovered between our small lives and this grandiose mass. We were approaching the shores of a new world. Behind us, at our feet, at the bottom of the sea of fog, we left the earth and all its miseries."

After crossing the Brèche de l'Olan at sunset, Henriot and Simond descended to Lavey, where they were joined by Camille Ravanel. The following day, they rose early and skinned up to the great rock window of Col de la Muande, its square-cut shadow cast across the glacier below where Valentine and I now stood.

"How moving are these arrivals at a col where one anxiously discovers the new half of the world," Henriot continued. "[We] had never been to the upper Valgaudemar, and never imagined that such a place could exist on this earth. It was a unique experience for us and, I'm sure, one rarely felt with such intensity." No doubt our encounter was less emotional than Henriot's in 1918. Set against the horrors of warfare, the transcendence of their mountain journey stood out with remarkable clarity. Looking back down the steep, sunlit slopes towards Valgaudemar, I wondered how our Parisian friends were coping on the icy snow. I could imagine that, removed from city life and having little mountain experience to draw upon, they too were living through intense moments that would remain etched in their memories for years to come.

Valentine skied first from the Brèche, making carefully calculated turns down the steep, exposed slope. We slid past Lac de la Muande on progressively better snow, skirting under the base of Les Rouies and climbing into the shade to find the concealed entrance to the colossal northwest couloir. Though I had been mountaineering in Vallon de la Lavey

on several occasions, the landscape felt magnified in winter. Our eyes roamed the valley, captivated by the immense, icy lines on adjacent peaks. The mysterious couloir above us disappeared out of sight between massive rock walls. Silvia broke trail on foot up the first gorge, which opened into a hanging bay of deep, cold snow. Back on skis, I took over, cutting a track with steep, delicate kick turns across the bay and up the ramp above. This led to a vast, hanging amphitheatre high on the face where we paused briefly to eat. Pressing on, we ascended a gully that ended at an exposed notch below the final rock face where we met the howling, northerly wind. We raced to pull on clothes as we considered the terrain above.

As I began up an icy runnel, carefully threading a line of snow across ridges of steep rubble, I abandoned any hope of skiing the line. Although the face was rarely steeper than 45°, it was horrendously exposed in these conditions, and the friable rock made using a rope impractical. Pulling over the top onto the plateau, we continued to the summit of Les Rouies, where the vast panorama of the wild Écrins interior unfolded before us. All around were peaks I still dreamed of: L'Olan, L'Ailefroide, Les Bans, a lifetime of climbing within this one massif.

From the summit, we bounced across the sastrugied glacial plateau, passed the Refuge du Pigeonnier and dropped into the avalanche gullies leading to Gioberney. Descending over 2000m, we encountered the full gamut of horrendous ski conditions and managed just a few graceful turns in the middle, between the icy crust above, and the sticky glue lower down. We finally reached the end of the road as the last orange glow drained out of the western horizon. Back at the car, I sent a message to the Parisian crew to check if they had made it down safely. They had, though their descent had turned into another full-day epic, arriving at their car at 3pm. No doubt, each of them left with a visceral appreciation for the complexity and beauty of venturing into the mountains in winter.

Above: Valentine skiing the south flank of Les Rouies. Left: Valentine and Silvia ascending the face above the northwest couloir of Les Rouies.

16 Refuge de Chalance (2550m)

Rif du Sap (1413m), Valgaudemar, Écrins, France

11 places (11 mattresses, 22 blankets)

Gas stove, no fireplace

Pipe by the refuge, stream by the path at c.2330m, or snowmelt

3h00. +1120m & 5km from Rif du Sap (Hiking – T3)

From Rif du Sap, follow the steep path N then NW, passing Cabane de la Muande to reach Refuge de Chalance. From there, a pleasant hike leads N to Col de Chalance (3012m), offering a stunning view into the Vallon de la Lavey.

Tour des Rouies – 21km, +2200m from Ref. de Chalance (Ski – 4.2)

This tour traverses a rugged and remote corner of the Écrins, rarely visited in winter. Crossing complex terrain, the journey requires stable snow conditions throughout. The route begins at Le Bourg (1175m), where the road is closed in winter. Follow the road for 3km to Rif du Sap. From here, ascend near the line of the summer path to Refuge de Chalance. This section crosses a steep, south-facing hillside that clears quickly of snow, so it can often be ascended on foot to c.1900m.

From Refuge de Chalance, head N to c.2730m, then traverse E, descending under the SSW spur of Aiguilles des Saffres. Ascend N and climb the S couloir of Col de la Muande (45°, 230m). In drier conditions, with an increased risk of rockfall in the couloir, an alternative is to ascend to the Brèche de la Muande, just W of the Aiguilles des Saffres, and ski or rappel the N side (45°, 40m). From Glacier de la Muande, ski N past Lac de la Muande. (To break the tour into 3 days, it is possible to continue down Vallon de la Lavey to spend the night at Refuge de la Lavey). Ascend NE past Lac des Rouies to Col de la Lavey. Descend 100m E, then traverse S (or further E to Passage de l'Âne) to reach Glacier des Rouies. Ascend this SW then S to the base of Les Rouies (3589m). Ascend the summit bastion via the ENE spur (40°, 80m). From the summit, descend the Glacier des Rouies, passing S of Col des Rouies. Ski 600m E, then descend SE (40°) and follow the combe under Pointe Duhamel. At 2750m, head SE towards Refuge du Pigeonnier, then navigate gullies SE then S to rejoin the road at Gioberney and follow this back to Le Bourg.

Les Rouies (3589m) via NW Couloir (900m, AD-, 50°)

This magnificent feature, splitting the NW face of Les Rouies, remains mostly hidden when near the mountain. The continuity and giant S-shape of this colossal couloir can only truly be appreciated from afar, yet views of this face are hard-won. This itinerary makes a great ski-alpinism route in winter and spring or can be completed on foot in early summer. Start by following the tour described above and locate the hidden entrance to the couloir at c.2670m, situated 500m S of Lac des Rouies. Climb the initial 150m couloir (40-45°) to reach a wide bowl on the L. Ascend this to the upper bowl at 3100m. From here, climb R to gain a shoulder near Pt 3364 below the final mixed face. Climb this via snow gullies and rocky ribs (45-50°, 150m) to emerge onto the plateau of Glacier des Rouies and rejoin the tour above. While the primary couloir makes an exceptional ski line, the upper 150m requires abundant spring snow to allow for a continuous descent.

Top: Silvia and Stefano in the loft of Refuge de Chalance.
Right: Valentine filling her water bottle. Opposite: Valentine by the refuge at sunrise.

CLUB ALPIN FRANÇA

VALGAUDEMAR

The Refuge de Chalance is one of the rare unguarded refuges in the Massif des Écrins. Situated high above the tranquil Valgaudemar valley, in the southwest of the range, the hut is perched in a cradle of flat ground, encircled by cliffs and sheltered from cosmopolitan life. Staying here feels like visiting a secular hermitage or a celestial observatory.

Constructed in 1970, the refuge lies midway between the Olan and Pigeonnier refuges and was originally intended as a stopover for hikers on the "Balcons du Valgo" trail. This route included several scrambling passages, such as the Col du Bâton, that was equipped with fixed cables. However, with the creation of the Parc National des Écrins in 1973, the management sought to limit the dispersion of hikers into such rugged "chamois" terrain, instead encouraging visitors to stick to a few classic routes.

Since then, the path is only maintained to Refuge de Chalance, which consequently has fewer visitors. Until 1999, the refuge was staffed in summer, with the tiny attic serving as the guardian's quarters. Today, volunteers from CAF Gap hike up to the hut once a fortnight during summer.

The refuge consists of a single room housing a cooking area, dining space and five beds on the ground level, with a mezzanine offering six additional beds above. Dry toilets and solar panels were installed in 2015. Until 2021, a pipe fed by meltwater from the slopes above the hut provided a reliable summer water supply, but increasingly frequent dry spells have since made this less dependable. The lozenge-shaped design of the building is characteristic of its era, resembling several refuges built in the Vanoise during the same period.

Above: Stefano approaching the Brèche de la Muande.
Top left: CAF sign on the hut exterior. Bottom left: Valentine, Silvia and Stefano inside the refuge.

Improvising in Belledone

Habert des Sabottes · Habert de la Pierre · Habert des Jarlons

As autumn approached, a week of bad weather brought the snowline down to tickle the 3000m summits of the western Alps. With our musician friends, Thomas and Dan, we left Chamonix and headed south. Their original plan had been to climb several classic routes around Chamonix and end their trip with an ascent of Mont Blanc. But with a tricky forecast of more snow and strong winds at altitude, we suggested exploring the Massif de Belledonne instead.

As this range isn't well known for its alpinistic heritage, it took some persuasion to convince Dan and Thomas to forgo the emblematic climbs of Chamonix and venture further afield. After a little research, the traverse of the Grand Pic de Belledonne, a long and committing route on the highest and most striking mountain in the range, seemed like a shrewd objective while the loftier peaks lay buried under fresh snow.

We arrived in the late morning, and while Valentine and I made a long detour to explore a valley to the west, we loaded Dan and Thomas with the ropes and sent them ahead towards Refuge Jean Collet and Habert de la Pierre to scout out a possible place to sleep.

From Col de Pré Long, Valentine and I climbed through woods and cow pastures, soon reaching Habert des Sabottes. We had a quick lunch there before continuing up towards Col de la Sitre, passing the iconic heart-shaped lake. Drawn in by a wisp of smoke rising from the hut, we went down to chat with the gaggle of elderly shepherds at the recently renovated Habert de la Sitre. They had just taken the sheep and their four guard dogs down for the season and were preparing to lock up the hut for winter. While discussing the notoriously protective Pyrenean Mountain Dogs, they agreed that poorly trained ones can pose a real menace to hikers, but could offer no better advice than to give the flock and its guardians

Left: Valentine near Mont Saint-Mury, with the Grand Pic de Belledonne behind. Above: Valentine and Thomas Encho on the Traversée des Trois Pics.

a wide berth. Like most shepherds in the region, they had lost four sheep to wolves that summer while their flock was grazing above Habert de la Pierre.

Passing over Col de la Sitre, we joined the rugged Haute Traversée de Belledonne, otherwise known as the Chemin des Bergers, or Shepherds' Trail, and followed it up to rejoin Dan and Thomas, who we found climbing on the overhanging crack above Habert de la Pierre.

Dan and Thomas are both world-class jazz pianists who have known each other for many years. We first met Dan, who lives in New York, on the way to the Dent Parrachée in the Haute Maurienne, where he was hiking with his father. Though based in the city, Dan had been climbing on and off since a young age. After that encounter, he got in touch and asked us to take him and Thomas, who had barely done any climbing, on an accelerated course in ridge climbing. Dan and Thomas made for an unusual rope team. Both deeply passionate about their craft, they chatted nonstop about music, concerts and art, occasionally breaking into song or miming a melody that had come to mind. When a mountain rescue helicopter hovered overhead, they paused, focusing on the rotor's thrum. "The resonance is D-flat," Thomas remarked, watching a rescue worker descending on the winch line towards the north face. "You can hear the harmonic series of the fundamental very clearly," Dan added, his head cocked to one side, "the fifth and seventh are particularly clear."

Thomas, based in Paris, is a sensitive and empathetic young guy, with the lean build of an urban runner rather than the scruffy bulk of the average hill climber. I immediately warmed to him, though I was apprehensive about how he might fare in the mountains.

My concerns proved unfounded, and despite his inexperience, he had a natural awareness of his body and surroundings and moved intuitively on rock. Both men had carefully maintained hands and were understandably wary about damaging them by jamming their fingers in cracks or pulling on tiny edges. Moreover, neither Valentine nor I wanted to be responsible for injured or mangled fingers. After all, their hands were their livelihood.

The following morning, we set off in the dark, following the rough trail leading to Lac Blanc and the north face of the Grand Pic de Belledonne. Though Dan and Thomas had climbed the Matterhorn together a couple of years earlier, they hadn't done any mountaineering since. In the half-light of dawn, they looked up with some uncertainty at the jagged ridge towering above us.

As I had never been to the Belledonne before, I looked at the map and saw summits that didn't even reach 3000m, so I assumed we'd be hiking up "montagne à vaches" and climbing just above the cow pastures. To the contrary, as we clambered up towards Col de la Balmette, the path disappeared into a chaos of blocks and shifting moraine. The Grand Pic appeared more like the rugged high mountains of the Swiss Valais than the rolling hills of nearby Beaufortain. Although it was only mid-September, it had already been bitterly cold. Streams had already frozen into cascading icefalls, and fresh snow dusted the remnants of the Glacier de Freydane.

At the time of the first traverse from the Grand Pic to Croix de Belledonne in 1890, the Freydane basin we had just ascended was still filled with a deep glacier. Joseph Couttet, then guardian of the newly built Refuge de la Pra in the valley to the west, was joined by Alphonse Derive and Aimé Millon to guide Georges Dodero and Maurice Gérard. The traverse had already resisted several parties, so success was far from assured. They crossed the Col de Freydane and, moving swiftly across the glacier, reached the Col de la Balmette, noting it took them just 31 minutes. With over 150m of ice now gone, the approach is now considerably more arduous.

While Valentine guided Thomas, Dan and I began up the north ridge, following a gloomy gully marked by steep, awkward steps. Above, we zigzagged up easy but loose ground until we reached a large memorial plaque and the base of some ancient steel cables anchored to the rock.

In 1858, several decades before the first traverse of the three peaks, Étienne Favier, Stéphane Chapuis, Alphonse Michel and Frédéric Perrin made the first ascent of the Grand Pic. They climbed via the northeast flank, an ambitious undertaking at that time. At Favier's suggestion, the local mountaineering club, La Société des Touristes du Dauphiné, installed the steel cables in 1875 to make the ascent safer and more accessible.

That same year, P. Gaufrès made the fourth ascent of the peak, guided by Favier. He described the upper section of the route as "difficult and tiring", requiring constant attention to hand and foot placements. Though "the holds you cling to are generally solid, it's wise to test them, for if one were to give way, the consequences could be dire ... the two cables, each twenty metres long, are useful in these near-vertical gullies; they ward off any serious danger, while also allowing you to climb quickly up these sections where you don't want to hang around."

Reaching the summit, I was struck by the complexity of the terrain we had climbed, and it was clear why cables had once been installed. This is the easiest way to the highest point in the massif, but even today descending this route would be a laborious and taxing endeavour. Looking ahead, the ridge stretching beyond us looked yet more precipitous and fearsome than it had from below. With a storm forecast for the evening, I wondered if we were heading, eyes wide open, into a trap.

In 1890, Couttet's team took an hour to breakfast on the summit of the Grand Pic. Gérard wrote, "Fog fills the valleys below, but the sky remains absolutely clear, and the view of the Oisans and Mont Blanc ranges is splendid." He continues, "The descent to the gap separating the Grand Pic from the Pic Central is on the south face via a couloir that is almost vertical but without any difficulties."

This couloir, just below our feet, was indeed steep and not particularly difficult. It even has good rock.

Top right: Valentine and Thomas Enhco on the Arête N, near the summit of the Grand Pic de Belledonne.
Bottom right and previous spread: Valentine and Thomas on the Traversée des Trois Pics.

Nowadays, it is sparsely equipped with bolts for a series of rappels or an exposed downclimb. However, the amphibolite, a dense metamorphic rock, offers neither spikes for rope belays nor cracks for modern protection. Equipped only with a hawser rope and hobnail boots, Couttet's team must have clambered down relying more on confidence and faith than any secure anchor that would hold a fall. At the base, they were "obliged to traverse almost horizontally towards the gap across slabby rock that offers only the tiniest of holds."

Following the line of the first ascensionists, after four rappels, we reached an insecure ledge far below the crest we aimed to climb. Adrift in a sea of exposed, slabby terrain we traversed and scampered up a loose gully to the ridgeline, where we were pleased to find solid rock.

On the first ascent, finding the ridge punctuated with awkward gendarmes and steep walls, Couttet's team downclimbed and traversed beneath the crest across abominably steep terrain. Crab-crawling across the southeast face, "across some fairly difficult rock", they passed underneath all the jagged crenellations and turrets that make the ridgeline such a magnificent climb.

The towers ahead looked alarming enough, and I could see why Couttet wanted to avoid them, yet the easiest line wove between them, only occasionally blocked by tricky steps. The first of these led directly up the crest of the second gendarme. With a stiff northerly breeze tugging at the rope, the exposure was wild. While the climbing was not technically difficult, grappling up an overhanging prow felt quite unnerving with 400m dropping on either side. Valentine was in the final stages of her guide training, and with Thomas inexperienced but quick to learn, he made an ideal partner to test her ability to lead in such demanding terrain.

Throughout the day, clouds had been gathering to the east, enveloping the mountains and approaching inexorably with a wall of rain and darkness. This confirmed the forecast and led a certain urgency to the afternoon. With meticulous routefinding and countless rope manoeuvres to safeguard our progress up, down and around towers, the ridge seemed to soak up time.

On the descent from the Pic Central, we came across the rusting remnants of another steel cable, this one installed in 1893. In the following year, two more sections were added, 60 and 80m long, running beneath the crest and across the southeast face to the base of the Grand Pic, nearly turning the traverse into a via ferrata. Upon their completion, Ernest Thorant, who had spearheaded these engineering efforts, claimed that "a strong climber could go from La Croix to the Grand Pic in 1h."

It was five o'clock by the time we finally reached La Croix de Belledonne and began down the easy path. As I turned the day over in my mind, what struck me most about the 1890 ascent was that they completed the first traverse from the Grand Pic to La Croix in just six hours. Navigating steep, delicate and unknown terrain with a five-man team, it seemed extremely rapid. We, on the other hand, had sacrificed speed for safety but nearly got caught in a storm for our troubles.

Less than 15 minutes later, clouds closed in and obscured the summit as we cantered down the west flank. The terrain around us was beautiful, rugged and mineral, but there was no time to pause and admire the view. Thankfully no pianist's fingers had been harmed during the climb. Yet, as predicted, just 20 minutes later, as we skirted back north towards the Col de Freydane, rain began to fall in waves, and the steep, slippery descent proved slow and arduous.

Left: Valentine and Thomas approaching the Pic Central de Belledonne, on the Traversée des Trois Pics.

17 Habert des Sabottes (1760m)

Col de Pré Long (1189m), Saint-Mury-Monteymond, Belledonne, France

6 places (3 mattresses, no blankets)

Woodstove in situ

Fountain outside the hut

2h00. +580m & 4km from Col de Pré Long (Hiking – T2)

From Col de Pré Long, follow the path SE to La Souille, then continue S to Col du Pré du Mollard. This point is also accessible by car via a forestry track. From there, take the path SE towards Refuge du Pré du Mollard, then at 1710m follow the track E to Habert des Sabottes. In winter, snow can block the route as low as Saint-Mury-Monteymond, in which case one can ski up the road and forestry tracks.

Petite Lance du Crozet (2355m) – 4km, +630m (Hiking – T3)

This minor summit gives a fine vantage point over Grenoble, the Chartreuse and Vercors. From Habert des Sabottes, follow the path SE to Mont St-Mury. Continue SE, passing above the turquoise crescent of Lac de la Grande Sitre. Follow the path to Col du Loup, then take the ridgeline NW to Petite Lance du Crozet (2hrs from Habert des Sabottes). From there, either retrace your steps or complete a loop by descending W to Lac du Crozet, then heading NW towards Pre Raymond, before traversing E back to Col du Pré du Mollard (5hrs from Habert des Sabottes).

The Habert des Sabottes nestles beside a large boulder on a grassy shelf high on the northern slopes of Mont Saint-Mury, in the heart of the Massif de Belledonne. The metal plaque fixed on the lintel above the entrance dates from its construction in 1920.

Built in a traditional alpine style, it features a masonry ground floor supporting a wooden loft space and a steeply pitched roof. Given that the hut is easily accessible and often busy, having a secondary option for overnight accommodation is recommended. There is space for a couple of tents in front of the hut.

The interior space is compact, dominated by a large table with seating for six. A rickety ladder leads to the sleeping loft, which also sleeps six. With limited insulation and a shaded, north-facing aspect, the hut stays cool in summer but can be chilly in winter, as it doesn't see the sun for a month on either side of the solstice.

Top: Valentine in the loft of Habert des Sabottes. Bottom: signage at the hut. Opposite top: departing from the hut. Opposite bottom: interior view of the hut.

1920

The habert stands within a small enclosed area that includes a large stone table with benches and a water fountain. While the hut has a wood-burning stove, wood is only supplied once a year. When this is exhausted, visitors can gather and split deadwood from the surrounding forest using the saw and axe provided.

The word "habert" is a regional term specific to the French Alps, particularly in the Belledonne and Chartreuse massifs, referring to a simple mountain shelter used by shepherds or alpine farmers during the summer grazing season. The Habert des Sabottes, owned by the commune of Saint-Mury-Monteymond, was built for this purpose, and for several generations, shepherds used it each summer to tend their flocks. Just above it stands the Refuge du Pré du Mollard, originally built in 1973 as a more substantial shepherd's hut and later converted in 2001 to accommodate hikers.

As transhumant farming, the seasonal migration from the valley to rugged mountainside grazing, becomes less viable, fewer pastoral shelters continue to serve their original function. While many have fallen into disuse, others have been renovated or repurposed. Some still support working shepherds, while others now serve mountain tourism, functioning either as staffed refuges or as simple, unmanned shelters. With the support of the commune, the Habert des Sabottes was renovated around 2002 by a team of volunteers and has since been kept open as a public shelter.

Since then, the commune has organised a work party each year on the first Sunday of October to maintain the hut and restock firewood before winter. Visitors are kindly reminded: "The wood gathered, chopped and stored by commune volunteers is to be used solely for heating in the stove, not for bonfires or barbecues."

Above: Valentine hiking above Mont Saint-Mury.
Top left: the loft in Habert des Sabottes. Top right: Dent de Crolles seen from the hut.
Bottom left: plaque dated 1920. Bottom right: the hut seen from the path above.

18 Habert de la Pierre (2030m)

Pré Marcel (1297m), Sainte Agnès, Belledonne, France

4 places (no mattresses or blankets)

No fireplace or permanent stove

Source 150m NW by the new shepherd's hut

3h00. +780m & 6km from Pré Marcel (Hiking – T3)

From the Col des Mouilles, take the minor road that climbs to Pré Marcel. From there, follow the marked path heading S, traversing the W flank of Jas Mouton. Before reaching Refuge Jean Collet, an exposed section is secured with a fixed cable. From the refuge, follow the path SE up to Habert de la Pierre.

Rocher de l'Homme (2763m) – 4km, +790m (Hiking – T3)

This hike leads you along wild and lonely paths to one of the highest peaks of the Belledonne. From Habert de la Pierre, head S then SW on a trail that traverses ledges on the exposed "Vire des Excellences", crossing the W flank of the Pointe des Excellences. Continue along this route as it descends to the N shore of Lac Blanc. From there, follow faint trails NE to Col de Roche Noire, then ascend the easy ESE ridge to reach the summit of Rocher de l'Homme.

Habert de la Pierre is a rudimentary stone shelter tucked beneath a massive gneiss block. Included here as a historical point of reference, it is a rare surviving example of the simple mountain refuges once common throughout the Alps. Functional yet austere, the hut has just three masonry walls, with the overhanging face of the gneiss block forming its rear. Inside, a wooden bunk bed for four people and a few shelves provide the bare essentials. The setting, high on the mountainside, is superb.

The Habert dates back to the 19th century and has been lovingly maintained across successive generations. Its wooden roof was restored around 2005, and is now clad in steel sheeting and reinforced with tensioned cables. Until the construction of the new pastoral hut, Chalet Denis Fort, in 2008, 150m to the northwest, the Habert de la Pierre was often reserved for shepherds during the summer months.

Though spartan, the Habert can serve as a contingency if the nearby Refuge Jean Collet is full. It may also be a deliberate choice for those resilient enough to spend a night here and glimpse the hardships once faced by those who eked out a living in these mountains.

Top: Valentine watching the sunset near Habert de la Pierre. Right: map of the region. Opposite: Habert de la Pierre and its gneiss block.

Just below the Habert de la Pierre stands the Refuge Jean Collet. Originally built in 1910, it was destroyed by a violent storm in 1926. A metal-framed replacement was erected two years later, then expanded in 1967. Further renovations in 1987 allowed the refuge to be staffed during the summer months. Outside this period, a few beds remain available.

Suzanne Ténot's 1919 study of human geography in the Massif de Belledonne offers a rare glimpse into the austere pastoral life that defined these valleys a hundred years ago. She described how, "above the last cereal fields on the slopes, between the forest and the bare rock zone of the summits, the areas available for grazing are very limited." The mountain pastures, she noted, were fragmented, interrupted by steep forests and the bare rock of the higher peaks, making alpine grazing a precarious and marginal pursuit.

In this unforgiving landscape, the shepherd's dwelling was equally rudimentary. Ténot wrote of the "habert" as "a simple stone or wooden structure where the shepherd comes to sleep, and which can shelter the flock in case of a storm." These shelters were scarce, especially on the massif's western slope, where communal herding practices meant that only one was needed for each basin. The Habert de la Pierre, she observed, "which appears to be the highest on the western side, reaches an altitude of 2020 metres."

She concluded that habitation in the Belledonne was "strictly adapted to necessity. There is no pursuit of elegance, nor even of comfort. It is clearly the work of a race of hardworking peasants, concerned above all with the struggle against an ungrateful land."

Above: Valentine inside Habert de la Pierre.
Top left: approaching the hut. Bottom left: departing from the hut.

Grand Pic de Belledonne (2997m) – Traversée des Trois Pics (350m, AD+, IV+)

Overview

Crossing the highest summit in the Massif de Belledonne, this is the finest alpine route in the region, linking the Grand Pic (2977m), Pic Central and Croix de Belledonne along a dramatic ridgeline. It offers sustained scrambling, airy exposure and a high-mountain atmosphere that feels surprisingly serious despite its modest altitude. The climb is easiest in summer and autumn, once the snow has melted from the approach. However, the route also makes a fine challenge in winter conditions.

Approach – 5hrs from Pré Marcel – 10km, +1500m

Several approaches are possible from the west, via the car parks at Pré Marcel, Pré Long or La Gorge. To shorten the ascent, climbers can stay overnight in Habert des Jarlons, Refuge Jean Collet or Habert de la Pierre. From Pré Marcel, follow the marked path heading S, traversing the W flank of Jas Mouton. Pass Refuge Jean Collet, then traverse S and head up to Lac Blanc. From there, follow the rough trail SE and up the N side of Glacier de Freydane. Scramble up E towards the S brèche of Col de la Balmette (2665m). Near the brèche, head R then L to reach the crest just above the notch at the base of the N ridge of the Grand Pic.

Col de la Balmette can also be reached from the east via Chalet du Rif Premier. From there, head NW past Roche Blanche, then ascend W up the scree and snow gully to reach the col. On the eastern side, an alternative, finer but more difficult alternative to reach the Grand Pic de Belledonne is via the E ridge, the Arête du Doigt (D-, IV+)

Grand Pic de Belledonne – Arête N (PD+, III)

This is the easiest route up and down the Grand Pic and an interesting climb in its own right. From Col de la Balmette, ascend R then L for 50m on the NW flank, then move L to the base of a loose gully. Climb this gully (step of III) to a fork, then take the L branch to reach a crest. Don't traverse horizontally L at this point (an exposed alternative); instead, head directly up broken terrain, trending L of the crest to find a memorial plaque at c.2850m, at the base of a steep buttress.

From the plaque, climb up L via the stepped slabs (old cables, III-), then move R to regain the crest. Follow this to a steep spire, then trend L and continue up via stepped buttresses (old cables) and a final gully to reach the summit of the Grand Pic de Belledonne (2hrs from Col de la Balmette).

Traverse to Pic Central de Belledonne (AD+, IV+)

From the summit, head SW towards Pic Central and scramble down 15m to a notch. Scramble S down a rocky gully for 10m to find a 2-bolt anchor on a ledge on the L (E), marking the start of the rappels. (It is possible to downclimb this gully on good but compact rock (III+, bolts), with a final 10m rappel to reach the base.)

1 – Rappel 30m down the gully to an anchor on the R (E, bolt + piton).
2 – Rappel 35m down the gully to a belay on the L (W, bolt + piton). Alternatively, rappel 30m and downclimb (II).
3 – Rappel 50m, trending L (W) across corners, passing several optional anchors (which need reinforcement), to a ledge at the base of a gully on the SE flank.

From this ledge, traverse 15m W and climb the gully (III) leading to Brèche Reynier. From there, climb along the NW side beneath Gendarme #1 (III-, bolt and pitons) to the notch beyond. Climb the second gendarme directly on the crest (IV+) for 25m to a ledge. Make an exposed traverse L (IV, piton), then continue on the SE side of Gendarme #2 to the notch beyond. Follow the crest (III) towards Gendarme #3, climbing directly up a steep step (IV). Continue easily across the crest of Gendarme #3 to the gap before Pic Central. Downclimb steeply (or rappel) on the SE side (III+) to the gap. Climb up the grooves beyond (III-) to reach Pic Central (3hrs from Grand Pic de Belledonne).

Traverse to Croix de Belledonne (AD-, III)

Descend SW, zigzagging down the crest (III-) to a notch and small tooth. Climb over this and descend past the old cables (III) to reach ledges on the W side. Ascend back to the crest of Gendarme #4, then scramble easily (II) down its SW flank to reach Brèche Duhamel. From there, ascend directly up the final fun section of crest (III) to reach the Croix de Belledonne (1hr from Pic Central).

Descent

From Croix de Belledonne, follow the path SW then N to the Col de Freydane. Descend N to Lac Blanc, then retrace the approach to return to the car park.

Equipment

Crampons and an ice axe are essential in the early season. A 50m rope plus a 50m tag line is required, or alternatively, a single 60m rope if you are comfortable downclimbing the SW couloir of the Grand Pic or making intermediate rappels.

Right: aerial view of the Grand Pic de Belledonne from the north, showing the line of the Arête Nord.
Following spread: aerial view from the east, showing the line of the Traversée des Trois Pics.

Grand Pic de Belledonne (2977m)
Pic Central
de Belledonne
Le Doigt
cables
II+
III-
III-
cables
c.2850m
II+
I/II
II
expo
III+
Col de la
Balmette (2665m)

Pic Central
de Belledonne
Croix de Belledonne (2926m)
Gendarme #4
Brèche Duhamel
III-
III-
III
II
III

Grand Pic de Belledonne (2977m)
Brèche Reynier
Gendarme #1
endarme #3
Gendarme #2
II
30m
35m
50m
II+
IV
IV+
III-
III

19 Habert des Jarlons (1890m)

Pré Marcel (1297m), Sainte Agnès, Belledonne, France

10 places (no mattresses or blankets)

Woodstove in situ

Source 100m W of the hut

2h00. +600m & 4km from Pré Marcel (Hiking – T2)

From the Col des Mouilles, take the small road that climbs to Pré Marcel. From there, various paths lead SE up to Habert des Jarlons.

Le Grand Replomb (2504m) – NW Couloir (Ski – 3.3)

Habert des Jarlons makes a fine base for ski mountaineering in the range between Rocher de l'Homme and Le Ferrouillet. The NW couloir of the Grand Replomb is a prominent feature, visible from afar. In winter, Habert des Jarlons can be approached from Col des Mouilles, though more reliable snow cover is often found by approaching via Habert de Crop, starting from the Pont de la Betta road above Prabert. From Habert des Jarlons, ascend S over Orionde and traverse S under La Rafour. Ascend S to reach the base of the NW couloir and climb this (40°) to a notch. From there, follow the final ridge to the summit (2hrs from Habert des Jarlons). From the summit, either ski back down the NW couloir, or continue to complete a circuit. An excellent loop (also possible in reverse) descends the SW couloir (passage at 45°), then climbs to the Passage de Jas Mounton. From there, ski the N couloir (40°) and the NW-facing combe below, before ascending back to Habert des Jarlons.

Built in 1990 through a communal effort by local volunteers and members of the "groupement pastoral" (the local farmers' association), the Habert des Jarlons serves both as a pastoral hut and an unguarded public shelter. Tucked into the alpine pastures beneath Orionde (2041m), a modest summit in the commune of Sainte-Agnès, it sits just above the treeline, overlooking the Grésivaudan Valley and the Chartreuse mountains. Facing northwest, the hut catches the last light of day and provides a peaceful base for hikers and ski tourers alike, ideally placed for a circular route through this quiet corner of the Massif de Belledonne.

Also known as Chalet Fernand Didier-Carnival, the hut was named after a local councillor who played a key role in its construction. Marc Bourgeat, one of the local farmers involved in the project, recalled: "There were about a dozen of us working over a couple of months." The site was chosen because "there had already been a hut there, though only a few stones remained. There was also a water source nearby."

Stone was collected on the mountain and transported to the site by mule and cart, while the rest of the materials, cement, timber and tools were airlifted by helicopter. The building reflects the region's pastoral architecture with a stone lower storey, a timber-framed upper level and a gabled roof.

The structure is divided into two parts. One section is locked and used as a storage space for pastoral equipment by the farmers who graze heifers on the surrounding pastures during summer. While the herders make regular visits, they rarely stay overnight.

The main room, however, remains open to the public year-round. It offers basic sleeping space for up to ten people on the wooden floor in the loft. There are no mattresses or blankets, so visitors must bring their own sleeping gear. A wood-burning stove provides heating and cooking, though users should bring firewood or gather deadwood from the nearby forest. While the roof is insulated, the solid stone walls retain little warmth, making the hut more suitable for summer use than for the colder months. A dry toilet is located 40m to the northeast.

Above: a young ibex above the path heading to Habert des Jarlons.
Top right: Habert des Jarlons at dusk. Bottom right: Valentine descending towards the hut.

Christine Baccon, a municipal representative from Sainte-Agnès, explained that while most pastoral huts are renovated with European subsidies, and therefore required to remain locked, others, like this one, are funded locally and can remain open. In autumn and winter, hikers and ski-tourers sometimes lack clear information about which shelters are accessible, and a small minority resort to forcing entry in search of shelter. The council observed that open huts were less prone to vandalism, and to avoid such problems, a section of the Habert des Jarlons was deliberately designed to stay open year-round.

Christine spoke candidly about the challenges facing pastoralism today. While several active shepherds remain in the area, with the pastoral group managing over 400 head of sheep and goats, the practice has become increasingly difficult to sustain in the current economic climate. The return of wolves to the area has also led to costly predation.

In response to growing tensions between tourism and pastoralism, especially regarding the aggressive Pyrenean Mountain Dogs that guard livestock, a new protocol has been introduced in the region: herders now notify local authorities when moving their flocks, enabling the commune and the tourist office to update hikers about livestock locations and help reduce potential conflicts.

The Habert des Jarlons remains deeply rooted in the region's pastoral heritage, and a work party from the commune of Sainte-Agnès and the "groupement pastoral" maintains the structure each year. Visitors are encouraged to contact the town hall in advance if planning an overnight stay, and to notify them if they find any damage.

Above: inside Habert des Jarlons.
Top left: Valentine and Thomas near the Croix de Belledonne, towards the end of the Traversée des Trois Pics.
Bottom left: Valentine descending towards Habert des Jarlons.

SAVOY PREALPS

Annecy
Thônes
22 / Chalet de Cotagne
Faverges
21 / Chalet du Charbon
Albertville
20 / Chalets Forestiers de la Balme
Frontenex
Aix-les-Bains
Chambéry
S A V O Y P R E A L
Rumilly
Seyssel
Cruseilles
La Balme-de-Sillingy
La Roche-sur-Foron
Saint-Pierre-en-Faucigny
Hauteville-Lompnes
Culoz
Belley
Yenne
La Motte-Servolex
Saint-Jean-d'Arvey
Saint-Pierre-d'Albigny
Challes-les-Eaux
Montmélian
Les Abrets
Pontcharra
Saint-Laurent-du-Pont
Voiron
Le Touvet
Crolles
Saint-Jean-de-Maurienne

Martigny
Cluses
Sallanches
Chamonix-Mont-Blanc
La Fouly
28 / Bivouac du Dolent
27 / Bivacco Cesare Fiorio
26 / Bivacco Gervasutti
PEN
29 / Bivacco Brédy
25 / Bivouac d'Estelette
Courmayeur
Morgex
30 / Bivacco Pansa
Aosta
GRAIAN ALPS
Val-Cenis

20 Chalets Forestiers de la Balme (1433m)

Raffort (719m), Frontenex, Bauges, France

Not intended for overnight use

Woodstove in situ

Source 50m N of huts

2h20. +700m & 4km from Raffort (Hiking – T2)

From Raffort above Clermont, follow the path W up through Le Grand Bois to reach the Chalets Forestiers de la Balme.

Le Grand Roc (1804m) North Ridge – 7km, +650m (Hiking – T3)

From the Chalets, head N to Col de la Fougère, then traverse N to Col du Haut du Four. From there, ascend the broad N ridge to reach Le Grand Roc (3hrs from the Chalets de la Balme). Return to Raffort by descending E via Chalets d'Orisan and Bois de Fournieux.

Mont Pécloz (2197m) SE Ridge – 6km, +980m (Hiking – T4)

From the Chalets, a pleasant hike leads N to Col de la Fougère and SW to Pointe de la Fougère, offering striking views of Mont Pécloz's NE face. From there, follow the easy (T2) but vertiginous crest S/SW to Mont d'Armenaz. A path then heads NW to Mont Pécloz, the 3rd highest summit of the Bauges (3h30 from the Chalets de la Balme). An exposed path (T4) descends W past Varapet into the Vallon de Bellevaux. Head ENE to Le Nant Fourchu, then ascend E to return to Col de la Fougère to complete the loop.

The Chalets Forestiers de la Balme are two stone huts perched at the treeline on the steep east face of the Pointe de la Fougère. Both have a charmingly rustic character, and their location makes them a convenient resting spot on long hikes through the Bauges.

The hillside above the Fournieux torrent, known as l'alpage de la Balme, has been inhabited since at least the 18th century. The "Mappe Sarde", the land registry of the Duchy of Savoy created in the 1730s, recorded a single pasture plot surrounded by extensive areas of uncultivated land. By 1870, seven small pastoral structures had established a precarious hold on this difficult terrain.

In 1903 and 1941, the state acquired over 200 hectares of this land to enable the Eaux et Forêts (Water and Forests administration, a precursor of the Office National des Forêts – ONF) to carry out reforestation work. The aim was to control the water flow of the Fournieux torrent, which during floods and heavy storms caused serious damage to crops and buildings in the valley below.

The project was ambitious: in addition to constructing a weir across the stream, they planned to plant seedlings over most of the hillside. The weir, now crossed on the approach path at 1150m, was built to slow the torrent's flow and promote sediment deposition. The plantations sought to recreate a forest environment to improve the soil's water absorption and help regulate stream levels.

In 1935, the chalets were built to house the reforestation teams, as well as the foremen and supervisors. Access trails were first cut at gentle gradients to allow materials to be hauled by mules and small carts. The inhospitable site was then levelled to create a large enough platform for the construction of buildings capable of accommodating the workers. Rectangular in plan, the chalets were constructed from limestone masonry and topped with gabled roofs, now clad in corrugated metal sheeting.

Previous spread: Marco Tibbetts on the north ridge of Le Grand Roc, with Mont Pécloz behind.
Above: chamois on the path near Col de la Fougère. Top right: Marco outside the southern Chalet Forestier de la Balme. Bottom right: inside the southern hut.

The commune of Cléry currently leases the southern building from the ONF. It remains open as a public shelter, although it is primarily intended for use by locals. This chalet is divided into three rooms by timber partitions, but there are no beds. We later learned from the ONF that, according to the rules of the Parc des Bauges, overnight stays are not officially permitted. There was firewood in stock, and deadwood was plentiful in the forest below. In winter, access is more challenging due to the steep trail and avalanche risk. Although both chalets have stoves, insulation is minimal.

The northern building is taller, with two floors, each containing a single room. It is usually kept locked and serves as a shepherd's shelter during the summer grazing season. As the flock is guarded by patous, large Pyrenean Mountain Dogs known for their protective and sometimes aggressive behaviour, hiking in this area is probably best enjoyed outside the summer months. When we visited in spring, the door had been inadvertently left open. The ground floor had a woodstove and four metal bed frames with old mattresses, and stairs led up to an empty upper floor.

In the 1930s, a million saplings were planted, including spruce, pine, larch, beech and maple. A tree nursery was even established near the chalets to provide saplings for future replanting. Despite this massive effort, most of the young trees were destroyed by avalanches before they could mature. A few survivors remain on protected spurs that weren't exposed to avalanches.

Today, this heavily forested corner of the Savoy Prealps is rich in biodiversity. Partway up the approach, one enters the protected area of the "Parc naturel régional du Massif des Bauges", created in 1995 and recognised in 2015 as a UNESCO Geopark. Covering 856km², the park includes a 52km² National Hunting and Wildlife Reserve (RNCFS des Bauges), created in 1913 on state-owned land to counter the dramatic decline in chamois populations. One of only ten such reserves in France, it is governed by ministerial decree and dedicated to conservation, scientific research, and the management of species such as chamois, roe deer, and mouflon, as well as black grouse, birds of prey and protected insects.

Above: Marco on the path below Chalet d'Orisan.
Opposite top left: Mont Blanc seen through the window of the chalet. Top right: a water source by the path below the Chalets de la Balme, at c.1250m.
Centre left: Marco gathering wild garlic (outside the park boundary). Centre right: view of the chalets from Le Grand Roc. Bottom left: the stove in the northern hut. Bottom right: wood reserves.

Mouflon, a wild sheep species native to Corsica and Sardinia, were introduced into the Massif des Bauges in 1954. Initially intended to enhance hunting opportunities and contribute to ecological studies, over time the mouflon population has become an integral part of the region's biodiversity.

In mid-May, I set off with my father to explore the eastern flanks of the Bauges, high above the busy Isère valley. We ambled up through the deciduous woodland richly scented with a carpet of wild garlic, winding between mature beech and a scattering of other species: oak, ash, sycamore, yew, Norway spruce, larch and whitebeam. The canopy, dense with fresh spring leaves, let through only occasional shafts of sunlight, keeping the temperature ideal for walking.

We reached the chalets in the late afternoon and cooked a simple meal in the smaller southern building, which appeared tidier and better maintained. A nearby spring provided fresh water, and lower down we had gathered a bunch of wild garlic leaves to add to our soup.

After an early dinner, we wandered up to the Col de la Fougère at sunset to take in the view, then descended just as night fell. Clouds had filled the horizon, obscuring Mont Blanc, but the Massif de la Lauzière glowed in the last of the evening light.

Since the 1930s, reforestation methods in avalanche-prone areas have evolved significantly, aided by modern equipment. Helicopters, unavailable at the time, now assist in transporting materials, and snow fences can be installed first to stabilise the snowpack before tree planting begins. Over the years, as avalanches have become less severe on these slopes, natural reforestation has gradually reclaimed the abandoned pastures around La Balme. Deciduous woodland is steadily returning, and the Fournieux torrent no longer seems to pose a threat to the valley below.

Above: Chalets Forestiers de la Balme.
Top left: Marco on the north ridge of Le Grand Roc, with Pointe d'Arcalod behind.
Bottom left: descending the path through Bois de Fournieux.

21 Chalet du Charbon (1639m)

Le Martinet (780m), Combe d'Ire, Doussard, Bauges, France

15 places (15 mattresses, 10 blankets)

Woodstoves in situ

Tap inside the chalet

3h00. +950m & 6km from Le Martinet (Hiking – T2)

From Le Martinet in the Combe d'Ire, hike SSW past the Cabane du Fayard. From there, several path options lead W, then N to the Chalet du Charbon. This is also the safest approach to the hut in winter.

2h30. +570m & 6km from Le Reposoir (Hiking – T3)

From Le Reposoir (1167m) above Bellecombe-en-Bauges, head NE over the Col de Bornette, then continue NE up a trail and a rock-cut passage to La Combe du Charbon. From there, head S over the Col des Platières to reach Chalet du Charbon.

Mont Trélod (2181m) via South Ridge of Banc Plat (Hiking – T5)

Mont Trélod is the 4th highest peak in the Massif des Bauges and is a popular hiking destination. While a good trail passes Le Planay and climbs gently to the summit, a more interesting route follows the beautiful narrow crest between Banc Plat and Dent des Portes. From Chalet du Charbon, head SW to Pointe de Banc Plat. From there, scramble along the crest heading S, following yellow markers over a few exposed, rocky sections at the start. Continue easily along the ridge, then, near the Dent des Portes, pass an exposed rock step on the L (E), then ascend steeply up the N side to reach the Dent. From there, head SE, then follow the path S along grassy crests and ascend SE to Mont Trélod.

The Montagne du Charbon, located in the northern part of the Trélod ridge in the Massif des Bauges, is characterised by a distinctive geological structure known as a perched syncline: a fold in which rock layers have been compressed into a trough-like formation.

This formation occurred during the Alpine orogeny, a mountain-building event that began around 65 million years ago with the collision of the African and Eurasian tectonic plates. The syncline's elevated position is a result of tectonic uplift and the erosion of the surrounding areas.

The name "Charbon" appears in manuscripts as early as the 12th century. The Montagne du Charbon most likely refers to the early small-scale exploitation of carbon-rich geological layers containing fossilised plant matter.

These sedimentary strata, rich in plant debris but not developed into high-grade coal, may have contained enough organic material to justify localised extraction, thus giving the mountain its name.

Nestled on the Montagne du Charbon, the first Chalet du Charbon was built in 1858 to shelter the shepherds tending their flocks in these high pastures. It was replaced by a new structure in 1875, which was later destroyed by heavy snowfall and only rebuilt in 1923 by a local carpenter, Pierre Demaison. After World War II, the building was eventually abandoned and fell into ruin.

In 2002, the parish of Doussard, which owns the land, restored the chalet. Since then, a local association, Tous au Charbon, has taken charge of its maintenance and improvements.

Above: the alarm bell in Chalet du Charbon.
Top right: Chalet du Charbon at sunrise. Bottom right: Marco on the path to La Croix du Roy, with Mont Trélod behind.

Rustic and functional, the Chalet du Charbon, also known as the Refuge du Charbon, is a large square building featuring a limestone masonry ground floor and a timber-framed upper floor, topped by a long gable roof covered in corrugated iron. Lean-tos have been added to the east and west sides for wood storage, gas cylinders and a dry toilet, while a fire escape has been constructed on the north side.

The interior originally consisted of a single room used for cooking and cheesemaking, with a semi-underground cellar at the rear and an attic sleeping area above.

Today, the ground floor features a communal kitchen and dining space equipped with a wood-burning stove for heating and cooking, a selection of cookware, and large wooden tables and benches. A sink occupies one corner, and in summer, spring water is piped into the kitchen taps. Behind the main dining area is a locked room for tools and supplies. The upper floor has been divided into three small rooms, with a total of 15 mattresses.

In the prairie to the north of Le Charbon lies Refuge de la Combe, a small sheep and goat farm known for its excellent organic cheese. In summer, it also offers accommodation and serves meals, making it a more comfortable alternative to the Chalet du Charbon.

Above left: wood stores at Chalet du Charbon. Above right: view from the window, with Pointe de la Sambuy in the background.
Opposite top: interior of Chalet du Charbon. Opposite bottom: Marco on the path above the Combe d'Ire.

In late spring, with winter snow still covering the high mountains, my father and I were looking for a low-altitude mountain region where the snowpack had already melted. We headed for the Chalet du Charbon, starting from the Combe d'Ire on the northern flanks of the Bauges. On the valley floor, a moist microclimate gives the woodland the appearance of a temperate rainforest: blankets of moss drape over the branches of beech, sycamore, spruce and silver fir.

That season, my father was concerned he hadn't found time to train before coming to the Alps and wasn't fit enough to climb hills, so we started slowly, inching our way up a steep forest track. We entered the Réserve Nationale de Chasse et de Faune Sauvage (RNCFS) des Bauges, a strictly regulated conservation area, and paused for lunch at the Cabane de Fayard, one of the many open shelters in the region, shaded beneath the dense woodland canopy. Before the reserve was established, the last brown bear in Haute-Savoie was hunted here in 1893 by the Comte de Boigne.

Above the treeline, marmots whistled and chamois scattered into the woods, but there were still no cows or sheep on the pastures, as the snow had only just departed and the meadows remained pale, with fresh grass and crocuses only just emerging.

We had expected the mountains to be quiet on a Friday in mid-May, but hadn't realised that VE Day was immediately followed by Ascension Thursday, so most people had taken Friday off to enjoy a long weekend. The mountains were humming with activity.

As we were nearing the refuge, a man and his ten-year-old son caught up with us, only to race ahead of my 70-year-old father and me. They feared there wouldn't be any space left in the refuge, and unfortunately, they were right. Arriving a few minutes before us, they triumphantly claimed the last two mattresses and seemed to have no scruples about forcing my elderly father to sleep outside.

As usual, we were carrying a tent just in case, and luckily, the meadows of the Montagne de Charbon lie just outside the regulated Bauges hunting reserve, making overnight bivouacking in a tent permissible.

At the refuge, we met Guy, a carpenter and one of the volunteers who helped maintain the hut. In his regular job, he had worked on the Refuge du Goûter, building its complex timber structure. Helping out at the Chalet du Charbon was like a busman's holiday for him.

Previous spread: sunrise on Mont Blanc, seen from the Montagne du Charbon.
Above: sunset on the south ridge of Pointe de Banc Plat, with Pointe d'Arcalod and Mont Trélod on the horizon. Right: Marco on the path to La Croix du Roy.

That afternoon, however, his eyes were wide with disbelief at the steady stream of hikers. By 3.30pm, he had already had to turn away a group of eight who'd come up intending to party. The hut was well over its intended capacity, and to add to the chaos, the tap inside had just snapped, requiring him to cut off the water supply.

Each year, Guy and others haul up around six tonnes of wood, donated by the ONF, in a convoy of 4×4 vehicles, stacking it to dry near the hut each spring and autumn. It's a quiet triumph of volunteer effort, but not without strain. The hut survives partly on a modest income of around €4000 per year, left by visitors in the overnight fee donation box. This helps cover transport costs for supplies, but maintaining the refuge always requires more.

Guy spoke candidly about the changes over the last fifteen years: more visitors, and a broader range of experience and expectations. Some arrive with loudspeakers, treating the hut like a party chalet. Although the refuge is maintained as a place of shelter and silence, that idea, he admits, now feels fragile.

Once the refuge came into the shade, I headed up to Pointe de Banc Plat, then scrambled along its spectacular southern crest to catch the evening light, and photograph Mont Blanc at sunset. On my return, I counted over a dozen tents clustered on the grassy ledges around the Refuge du Charbon, with a hive of activity surrounding the hut.

Following Guy's advice, we had pitched our tent further around the hillside to the north, near La Croix du Roy (1711m), a large cross marking a spectacular viewpoint. In 1967, the bell tower of Doussard was struck by lightning and burned down. The metal cross survived intact, so the locals carried it up in pieces and reassembled it at its current spot.

Tucked out of sight of the hut on the northern fingers of the Montagne du Charbon, we actually had a far grander outlook than the chalet itself. Aside from the lumpy ground, trodden by centuries of cattle, it was one of the finest camping spots we'd found. We gazed down over Lake Annecy, with La Tournette's craggy summit rising above and Mont Blanc looming in the distance. I was glad to share it with my father, who had handled the ascent well despite his worries, and I was all the more thankful that we'd found a quiet corner away from the crowds.

Above left: Guy at the Chalet du Charbon. Above right: a waxing crescent moon seen from the Montagne du Charbon.
Top left: Marco making dinner at our camp by La Croix du Roy. Bottom left: overlooking Lac d'Annecy at dusk.

22 Chalet de Cotagne (1469m)

Chamossière (770m), Thônes, France

6 places (6 mattresses, 6 blankets)

Woodstove in situ

Spring beside the hut, but it may dry out in summer

2h00. +700m & 6km from Chamossière (Hiking – T2)

Walk from the bus station in Thônes, or park at c.770m beside the road above Chamossière. From the road, ascend W via a steep concrete track to 866m, where a trail branches NW. Follow the "Chemin de Cotagne" NW, then SSW to reach the small clearing and hut at 1469m.

Rocher de Belchamp (1860m) 1h00. +390m & 1.5km (Hiking – T3)

The northern summit of the Montagne de Cotagne is called Rocher de Belchamp. From the hut, follow the small unmarked path S up the ridge. At c.1740m, the rough trail traverses onto the E flank before ascending to the summit.

La Tournette (2350m) via Montagne de Cotagne (Hiking – T5)

La Tournette can be reached by an exposed path traversing Montagne de Cotagne. From Chalet de Cotagne, follow the small unmarked path leading S up the ridge to the path junction at c.1740m. Take the rough path leading R (SSW) along the W flank of Rocher de Belchamp to reach the col (c.1760m) between Rocher de Belchamp and Montagne de Cotagne. From here to Rosairy the terrain is steep and the correct route is hard to follow. It is possible, but more exposed and technical, to follow near the crest to reach Montagne de Cotagne. Usually, from the col, head c.100m down the E-facing couloir, then traverse at c.1660m along the E flank. Follow an exposed path (T5, faded red paint marks) across the mountainside at this level, then ascend diagonally beneath Montagne de Cotagne's main summit (1917m). Join the exposed path (T5) that leads towards Rosairy, and descend this SE to find the break in the cliff band on the SE flank. Follow the path down through this exposed terrain to join the main trail leading to Col des Frêtes de Rosairy. From there, head S and ascend the NE flank of La Tournette (T3) to reach the summit (5hrs from Chalet de Cotagne).

The Montagne de Cotagne forms the long northern spur of La Tournette, the iconic summit that towers above Lac d'Annecy. Perched on the eastern flank of this ridge stands the Chalet de Cotagne, overlooking the town of Thônes, the capital of Reblochon cheese production. The valley marks the boundary between the Massif des Bornes and the Chaîne des Aravis, two distinctive limestone ranges west of Mont Blanc.

Paul Mathevon, a long-time local resident and hunter, reminisced about the former hut: "When I was a kid, we used to sleep in the old chalet. The [pasture at] Prés de Cotagne was much larger back then because animals were still grazing up there … it was a proper old-style chalet, built the traditional way, with a stone base and wooden upper floor. There was a stable for the animals, and Joseph Donzel did the milking right there in the chalet."

Robert Donzel, now in his mid-eighties, recalled that his uncle Joseph kept a small herd of cows on Cotagne and another on a nearby hill. However, in the mid-1950s, Joseph gave up herding. The chalet quickly deteriorated, and in 1961 the roof collapsed and the building fell down.

Paul Mathevon remembered: "Docteur Bleyon, who was the mayor at the time [1959-1965], had the metal hut built." This structure, the current Chalet de Cotagne, was installed on the site of the ruined stone building. The new shelter was basic and functional: an uninsulated, prefabricated hut made of galvanised steel, set on a concrete base. Erected by the local municipality, it served as a simple open refuge for hikers and hunters seeking shelter on the ridge.

The design, also seen in the high mountains above Saas Grund at the Cresta-Biwak, was developed by the French manufacturer Ferdinand Fillod. These structures were strong, lightweight, quick to assemble and easily adapted to a range of environments. Built from cold-formed arched steel struts with ribbed mid-steel panels, they could be extended to almost any size by adding more segments.

Above: Valentine hiking through the forest towards Chalet de Cotagne.
Top right: arriving at Chalet de Cotagne. Bottom right: interior view of the hut.

Mass-produced before the Second World War, Fillod huts were widely deployed by the French army along the Maginot Line. Their rounded shape cast less conspicuous shadows, providing better protection from aerial reconnaissance. After the war, they demonstrated their versatility again, serving as shelters in both Antarctic research stations and tropical outposts.

Jean-Bernard Challamel, Mayor of Thônes from 1983 to 2014, recalled: "About thirty years ago, the shelter at Cotagne was used by a shepherd who grazed a flock up there, but afterwards it was mainly used by local hunters."

Once the pastures were abandoned, natural reforestation gradually reclaimed much of the mountainside. Aerial photographs from 1952 show over seven hectares of open ground on the mountaintop, along with a similar area of sparse woodland. Over the sixty years since Joseph Donzel maintained the land, the prairies have all but vanished. Today, only a small one-hectare clearing surrounds the metal shelter.

The same pattern can be seen across the Alps: over the last century, mountain regions have experienced a steady decline in pastoral farming. Demographic changes have driven this shift, with young people moving away from the arduous, low-income work of mountain herding. Economic policies, modernisation and mechanisation have also played a role, favouring valley agriculture over high-altitude grazing. This has led to widespread, rapid reforestation as nature reclaims spaces once cleared and maintained by herders.

Nicolas Genans-Boiteux, president of the local hunting association, confirmed: "There are quite a few in our community who stay at the hut and help maintain it a bit, trimming the bushes and trees in the clearing to keep them from encroaching too much."

Other than local hunters, the hut is mainly used by hikers. Despite its modest comforts, it is visited by enthusiasts throughout the year, as the well-thumbed visitors' book attests. Facing east, the Chalet de Cotagne catches the sunrise through the front door. Perched high above the valley, it offers sweeping views of the Aravis mountains, which become even more striking higher up the ridge.

Above left: interior view of the hut. Above right: view from the window towards Mont Charvin.
Top left: hiking through the forest towards the hut. Bottom left: Chalet de Cotagne and the remaining pasture.

The hut has two small rooms with a metal partition. The dining area contains a sturdy table, two benches, a wood stove and various cooking utensils, creating a warm, functional space. Beyond this lies the sleeping quarters: two sets of bunk beds, with accommodation for six.

Despite wear and tear, the interior retains a rustic charm. The peeling paint and timeworn stove add character. A humorous sketch warns: "Beware of the nocturnal smoker." With its rusty hearth and leaky flue, one apparently risks being cured like a fine Scottish salmon. Outside, a wooden basin is fed by a water pipe, though in early September the flow had slowed to a trickle. The supply is likely more generous earlier in the summer.

Valentine and I visited for a reconnaissance on a bright September day. With paragliders in tow and the day warming rapidly, we hoped to fly back down to the valley. However, the slope below the hut was far more overgrown than we expected, its dense bushes posing a serious risk of snagging our lines during takeoff.

Our only viable option was a narrow gangway between trees in the top corner of the clearing, a space barely wide enough to launch and requiring a tricky takeoff manoeuvre that involved brushing the wing against the trees on either side before getting up to speed.

With some trepidation, I launched first. Initially, the wing didn't provide enough lift, and I ran out of usable runway before it fully inflated. Unable to stop, I had to leap over a boulder before finally getting airborne, just clearing trees.

Once away from the hut, warm air rising up the hillside provided steady lift, and I caught a few thermals, circling back to watch Valentine prepare for her takeoff. With a smaller wing, and being lighter and more experienced, she executed a much cleaner launch. With clouds thickening rapidly over La Tournette and a storm forecast for the afternoon, we made a swift descent to the safety of the valley.

Above: sunset view from Pointe de Talamarche, with Montagne de Cotagne on the left and La Tournette in the centre.
Top right: Suzanna Walker and Amy Shanton Foo at sunrise near Pointe de Talamarche, with Montagne de Cotagne behind.
Bottom right: taking off from Chalet de Cotagne (Photo by Valentine Fabre).

GRAIAN ALPS

Sallanches
Chamonix-Mont-Blanc
28 / Bivouac du Dolent
La Fouly
27 / Bivacco Cesare Fiorio
26 / Bivacco Gervasutti
29 / Bivacco Brédy
25 / Bivouac d'Estelette
Courmayeur
30 / Bivacco Pansa
Morgex
Ugine
Albertville
G R A I A N A L
Val-Cenis
23 / Rifugio Santa Maria
Modane
11 / Bivacco Colle del Sommeiller
14 / Bivacco Hannibal
Susa
13 / Bivacco Walter Blais

Breuil-Cervinia
N N I N E A L P S
32 / Bivacco Tzan
Valtournenche
31 / Bivacco Rivolta
Gressoney-
La-Trinité
Saint-Vincent
Pont-Saint-
Martin
Biella
24 / Bivacco Beltrando
S
Ivrea
Courgnè
Castellamonte
Strambino
Alice Castello
Santhià
Caluso
Cigliano
Mathi
Ciriè
Verolengo
Crescentino
Settimo
Torinese
Venaria Reale

Rocciamelone and the Origins of Mountaineering

Rifugio Santa Maria

On a bright day in early November, Valentine and I drove over the Col de Montgenèvre towards Rocciamelone. At 3538m, this broad pyramid towers above the Piemontese town of Susa (503m). For centuries, it was, somewhat surprisingly, thought to be the highest mountain in the Alps. We planned to spend the night in the tiny chapel-refuge perched on the summit. As it was All Saints Day, I wondered if we might have to jostle to find a bunk space up there. The sky was clear, but temperatures at the top were forecast to be well below freezing. I was curious to experience the ascent in conditions that might resemble those of the medieval period, when the mountain was first climbed and summers were markedly cooler.

As we cruised down the Susa valley, Rocciamelone appeared to have only a light dusting of powder near the top. That perspective turned out to be vastly misleading, and we later found drifts of snow, metres deep. The exposed hiking path to the summit had been transformed into a steep snow slope punctuated by rocky buttresses. If there had been any solid rock to attach oneself to, the ascent would have felt far more secure with a rope.

As legend has it, in the mid-14th century, a local nobleman, Bonifacio Rotario D'Asti, of Roero near Asti, was captured by the Turks while crusading in Anatolia. During his imprisonment, he vowed that if he regained his freedom, he would have a chapel built in honour of the Virgin on the highest peak in the Alps. Upon his return, he settled near Susa and made the ascent of Rocciamelone on 1 September 1358. Although this is the first confirmed ascent, records suggest the mountain may have been climbed much earlier to make offerings to Jupiter. On Bonifacio's ascent, a small chapel was installed in the summit rocks and a gilt bronze triptych was erected on the site. In subsequent years, this

Left: Valentine on the south ridge of Rocciamelone. Above: an old ibex on Rocciamelone.

perilous climb became a popular pilgrimage for devout Christians from the surrounding valleys.

As the climate cooled after the 14th century, Europe entered what is now known as the Little Ice Age. During this period, a major glacier advanced over the north side of Rocciamelone, gradually engulfing the summit. The original chapel became enveloped in several metres of ice (only re-emerging in 1820), and a small wooden sanctuary was constructed slightly lower. This was rebuilt many times until the last one burnt down in 1913. After the First World War, the current masonry building was erected, housing a chapel, sacristy and a small public refuge.

Historical references to Rocciamelone are scarce, but Jacques de Villamont, a French noble at the court of Henri III, left a detailed description of his ascent. In 1588, Villamont embarked on a three-year tour of the Mediterranean, travelling from Paris to Lyon, then on to Turin, Rome, Naples, and all the way to Palestine and Egypt. Following passage through the plague-ridden Maurienne valley and across the Col du Mont-Cenis, he was descending Valle di Susa when he was arrested and required to quarantine in Novalesa before entering Turin. During his wait, he recorded that "several of the inhabitants suggested I go to Notre-Dame-de-Rochemelon, which is a small chapel on the top of a mountain bearing the same name." Intrigued, Villamont set out to climb the mountain "to satisfy my mind with something that is so unusual and novel." He hired two guides to show the way and carry food for two days. Halfway up, Villamont became so weary he could not continue. After spending the night in a cheesemaker's cottage, they resumed their climb, now accompanied by the cheesemaker's son. "Immediately at daybreak, we continued the steep ascent of the mountain, which seemed much more difficult than at the beginning."

Nowadays, you can drive up to this point, 1500m above the valley floor. Winding up the tortuous road, Valentine and I passed hectares of carbonised larch woodland, decimated in the forest fire that swept across Valle di Susa in 2017. We parked below La Riposa, where, according to legend, Bonifacio Rotario D'Asti rested ("reposed") before continuing his ascent in 1358. Lazing in the midday sun, we ate some bread and cheese and began to pack our bags. As I squeezed my sleeping bag into my pack, my heart sank as I realised I'd forgotten our camping mattresses. Until 2019, the shelter on the summit had been equipped with bedding, but due to overcrowding and mistreatment, they were removed. The thought of sleeping on wooden boards at 3500m on a freezing November night made me shudder. Searching in the van, I wondered if the gritty floor mats might be of any use. I then remembered the old woollen army blankets Valentine kept under the seats to cover dirty bikes. I rolled one up and strapped it to the outside of my colossal pack.

As we were about to set off, the guardian of nearby Rifugio Cà D'Asti passed on his way down after some maintenance work. He informed us that, due to vandalism, the only winter shelter at his refuge was a bare stone room. There was no longer a bunk room available, let alone mattresses or blankets. With a grin and a hint of sarcasm, he added: "You can sleep on the table if you want." Cà D'Asti, the highest staffed refuge for ascending the mountain in summer, was supposedly the site where Bonifacio Rotario D'Asti's team erected a small wall and bivouacked on their 1358 ascent: the Casa ["house"] D'Asti. Laden like mules, Valentine and I set off slowly up the hill, passing a few people who had turned back when conditions became too difficult.

We reached the snow line at 3000m, where the terrain steepened. The muddy path yielded to a mineral, high-mountain landscape. Vast slabs of crumbling rock were covered with patches of névé, and an old male ibex watched us from the crest above, his massive horns cutting the skyline. We continued slowly upwards as he observed us nonchalantly before ambling down and disappearing from view. A sturdy breeze picked up as we gained the fore-summit, marked by an obelisk and a cross, the Croce di Ferro at 3306m. With snow covering most of the terrain above, it became clear that no one had ventured higher since the recent storms. We donned crampons and investigated further. The broad ledge where the path usually continues was buried, leaving only a steep snow slope that drew the eye down the precipitous southeast face. To avoid that treacherous route, we gained the south ridge, a narrow rock crest draped in névé and exposed to the frigid westerly wind. With the sun approaching the horizon, we carefully ascended, teetering on a tightrope high above the usual path. Scrambling along the crest, we battled against a rising gale as we approached the final rocky bastion below the summit.

"When we had passed three leagues in height," writes Villamont, "I had to attach iron claws to my hands and feet, to climb upwards, and also for fear of slipping down the precipices, which threatened us with a horrible death. It was then that the aid of the guides became crucial. Without them, I would not have readily wished to risk my life. As I gradually approached the top of the mountain, in the middle region of the air, I was immediately seized by an unbearable cold, so that, changing colour and being completely exhausted and distressed, I was forced to drop to the ground to take some rest. The guides, accustomed to this work, made me drink a little wine to give me the courage to continue our journey. Finally, with their help, I arrived near the summit of the mountain, where I believe I have never endured more violent cold, nor made such a perilous journey. For one must then ascend, as if by a ladder, climbing the mountain with the iron claws attached to one's hands and feet ... one cannot expect anything but annihilation for those who slip."

I could well understand why Villamont found the ascent so taxing in the cooler climate of the 16th century. Even today, the summit is only snow-free for a brief period, and most ascents occur during the hottest months of the year. In Villamont's time, however, the final vertiginous bastion would have been a steep climb over shattered rock, often covered in patches of snow and ice, even at the height of summer. Furthermore, the iron claws used by Villamont and his guides were far less secure than the modern crampons we were wearing. The guides must have worked tirelessly to ensure his safety with the rope, as the rock remains loose even after 500 years of ascents and offers few reliable anchors.

Top right: Valentine on the south ridge of Rocciamelone at sunset. Bottom right: descending Rocciamelone's normal route – slightly more than a T4 hike in snowy conditions!

As Valentine and I climbed the final bastion of the south ridge, we could see flecks of red and white paint on protruding rocks to our right, marking the normal way. In several places, rows of iron rods were visible, supporting the rough masonry that holds the modern path in place. In summer conditions, with steps excavated into the rock and fixed hawser ropes for added security, the ascent of Rocciamelone is a far cry from the steep, icy rubble faced by Bonifacio and Villamont in the 14th and 16th centuries.

Jacques de Villamont wrote extensively about his travels, leaving us with the precious account of his ascent in his *Les Voyages du Seigneur de Villamont*, published in 1595. This book was one of the first to give practical guidance for travellers, as well as descriptions of the interesting sites he encountered. It proved very successful and went through numerous editions. Five centuries later, copies can still be found for sale. Written in Middle French, the small vellum-bound book is somewhat challenging to read. Much like old English texts, a soft "s" is printed in a character similar to an "f". I mouthed out the words, feeling like the lisp I once had as a child was returning to haunt me.

"Eftant doncques paruenu iufques au fommet l'entray en la Chappelle pour faire ma priere," continues Villamont: "Having therefore reached the summit, I entered the chapel to say my prayer, and immediately afterwards I went out, looking out over a large frozen lake, which is towards the country of the Grisons: then turning my head to the other side, I looked at the blades of mountains, both from Savoy and Dauphiné, still covered in their white hats, even though it was the month of August. And although all these mountains are very high, compared to the mountain where I was, they look small. Then when I set my eyes on the peaceful lands of Piedmont and Lombardy, I suddenly forgot all my past labours, and felt my soul filled with incredible joy: And in this joy, I longed to contemplate them more closely once descended from the mountain, to enjoy them more fully."

By 1588, Rocciamelone was not the only mountain in the Alps with a chapel near the summit, though it was and remains the highest. The people of this frontier region were pious folk, and in the surrounding valleys chapels had been constructed on other peaks. In 1446, the Chapelle de Notre-Dame was built on the summit of Pointe de Tierce at 2973m, while Chapelle Notre-Dame-des-Sept-Douleurs, also from the 1400s, was erected on Mont Thabor at 3171m. Ordinary people would make the pilgrimage, seeking commune with higher forces, both literally and metaphorically. As a modern panel in Rifugio Cà D'Asti still suggests: "With every step towards the top, you feel the presence of God ever closer."

The standard historical narrative suggests that before the Enlightenment and the Romantic movements of the 18th century, most people feared the mountains and disliked barren, inhospitable landscapes. Religious motivations were thought to be among the few reasons people might head into high mountain regions. As Macfarlane writes in his 2003 classic *Mountains of the Mind*, "Three centuries ago, risking one's life to climb a mountain would have been considered tantamount to lunacy. The notion barely existed, indeed, that wild landscape might hold any sort of appeal."

Before 1700, the story is more nuanced. In Europe, our understanding of people's relationship to the Alpine landscape is extrapolated from so few lines of evidence that our conclusions are inevitably skewed. The historical narrative is shaped by the written documents, maps and paintings left to us by the urban elite: the nobles, gentry, military leaders and clergymen. Although these records are fragmentary and represent only one segment of society, they have led modern historians to conclude that the Alps held little secular appeal and remained largely "terra incognita" throughout this period.

Not only are primary documents so scarce that our conclusions are little more than conjecture, but most writing left by the ruling classes is unlikely to correlate with the experience and knowledge of the broader population. Then, as now, chamois hunters and sheep herders crossed the high mountains, leaving scant record of their passing. Archaeological findings attest to rudimentary shelters in high alpine terrain stretching back millennia, suggesting that through oral tradition passed on by hunters, herders and smugglers, all but the highest glacial passes were already intimately known landscapes well before the Renaissance.

Left: sunrise at Rifugio Santa Maria on Rocciamelone.

By the early days of "aristocratic" mountaineering, of which Villamont is a precursor, the chamois and crystal hunters became the de facto mountain guides, possessing the most intimate knowledge of the terrain and how to survive in the high mountains.

Villamont's curious account, albeit recorded by a member of the ruling class, among other scraps of evidence from writers such as Benoit Marti, Conrad Gesner and Petrarch, suggest a different narrative is plausible: long before the Romantic movement, people were sensitive to the beauty of a mountaintop view. Far from a universal belief that the high mountains were ghastly, barren realms inhabited by dragons and malevolent forces, some people climbed mountains out of curiosity even in the 1500s. The fact that Villamont was able to find guides and crampons in the remote corner of Valle di Susa suggests that such ascents were not as unusual as we might assume.

Although Rocciamelone was a sacred site, Villamont's writing suggests that his ascent was driven not by religious devotion like the pilgrims, but by sheer curiosity to experience new sensations and perspectives, something "so unusual and novel". While his secular curiosity may have been uncommon, it was probably not unique. However, as it stands, we have scant written records of people in the high mountains during this period, regardless of their motivations, whether secular or religious.

Upon reaching the summit, Villamont briefly mentions the religious structure before launching into an extended description of the sublime view: "I suddenly forgot all my past labours, and felt my soul filled with incredible joy." His account, focused on the secular experience, remains the only known first-hand source from this period of anyone reaching a mountain peak over 3000m. Villamont's admiration for the landscape, along with the emotional and physical perspective granted by the elevated viewpoint, offers a fascinating precursor to modern mountain literature. Yet, like many of us at the top of a difficult climb, his thoughts were likely dominated by a more immediate concern – how to

Above: Valentine climbing the final buttress of the south ridge. Right: leaving Rifugio Santa Maria. Previous spread: Valentine on the south ridge at sunset.

get down from such a perilous place: "I longed to contemplate them more closely once descended from the mountain." I suspect that most early aristocratic travellers were repelled by this barren landscape simply because they had never bothered to climb a mountain themselves.

As the sun touched the distant horizon beside the Meije, Valentine and I were still 50m below the summit, blocked by a steep crumbling buttress. I had misread the easiest line and was now looking forlornly over the gully to our right, where stretches of fixed rope emerged from the snow. Unable to traverse across steep, icy slabs, we continued up an awkward chimney of loose rock. Glancing up as the light faded, a chough was hovering just above us, stationary on the updraft and looking down at us quizzically.

Upon reaching the summit, I didn't experience the euphoric visions Villamont described, for the sun had set 20 minutes earlier. Dusk was closing in, and the wind was raging. The Cappella-Rifugio Santa Maria was an impressive structure, but it was half-buried in snow. Fortunately, the top half of its metal door was still visible, allowing us to crawl through and drop into the frigid shelter. It was barely warmer than outside, but at least we were out of the wind. The small room was austere, yet after an ascent far more demanding than expected, it was a deeply appreciated refuge.

The wood-lined interior was sparse, with simple bunks arranged on two levels. A wooden table and benches completed the furnishings. The rest of the space was bare, save a few votive offerings: crosses, rosary beads and a couple of transparent plastic Madonnas filled with holy water from Lourdes. After gathering snow from the doorway and melting it for water, we had cooked and eaten by 7pm. Outside, a canopy of stars mirrored the bright tapestry of lights spreading across the plains beyond Turin. Between these, I could just make out the inky silhouettes of mountains. However, it was too cold for reverie. Like Villamont, I longed to contemplate all this once we got down safely. At that moment, there was little else to do but brace for the rigours of sleeping on wooden boards, softened only by the thin wool blanket I had carried and a few scraps of leftover foam. It was going to be a long night.

23 Rifugio Santa Maria (3538m)

La Riposa (2050m), Susa, Piemonte, Italy

12 places (no mattresses or blankets)

No fireplace or permanent stove

Melt snow if clean patches remain

There is a basic toilet by the path, just below the summit

4h00. +1500m & 5km from La Riposa (Hiking – T4)

From the car park below La Riposa, follow the path N up past Rifugio Cà d'Asti (2854m) and continue up to foresummit by Croce di Ferro (3306m). Traverse N across the SE flank, then zigzag steeply up the final bastion, equipped with ropes, to reach the summit platform and the chapel-refuge. Outside of dry mid-summer conditions, the difficulties can increase dramatically.

Traverse of Rocciamelone 10h00. +2000m + 23km (Hiking – T4)

This long hike can be split into two days, staying at any of the refuges en route. From Lago di Malciaussia (1805m) above Usseglio, follow the path to W Rifugio Tazzetti (2642m). Continue WNW on a path to Colle della Resta (3183m) then head SE across the flat glacier. Ascend the NW ridge via a faint path and easy scrambling to reach the summit (6hrs). Descend the normal route to Rifugio Cà d'Asti (2854m), then follow the path E to Capanna Ravetto, crossing Colle della Croce di Ferro (2558m) before heading back N to Lago Malciaussia.

Rocciamelone (3538m) via ESE Ridge (PD, III+)

The ESE ridge features long sections of enjoyable, easy scrambling, with generally good rock on the few steeper passages of III/III+. From La Riposa, follow the path N to the junction (2800m) just below Rifugio Cà d'Asti, then take the trail descending E. After the path bends SE at c.2570m, head NE up the hillside to the plateau at 2750m. Continue NE up the loose, rocky gully to Passo Brillet (2952m). Follow the narrow crest NW to reach I Brillet (3045m). After a short descent, reach the base of the three steep steps of Rocce Rosse (II, II then III+). Climb directly up these, bypassing some overhangs on the R, to reach the top of the buttress. From here, the rock quality deteriorates, but the route continues more easily along the crest all the way to Rocciamelone (6hrs from La Riposa).

Rifugio Santa Maria is built on the south side of Rocciamelone's summit at 3538m. This sturdy, squat stone building is clad in stainless steel tiles. It houses a central chapel with a sacristy and accommodation for priests in the west wing, while the east wing contains a small unguarded refuge. Replacing previous wooden chapels built over the preceding 500 years, the current building was designed by the architect Natale Reviglio, constructed between 1917 and 1923, and then renovated in the 1980s. In recent years, due to overcrowding and vandalism, mattresses and blankets were removed from the public refuge. The spartan, wood-lined room now contains only a table, benches and wooden sleeping platforms.

Above, right and opposite top: Rifugio Santa Maria on Rocciamelone.
Opposite bottom: Rocciamelone's southeast face. The left line indicates the normal route, traversing below the south ridge, while the right line shows the east-southeast ridge.

Rocciamelone (3538m)
I Brillet (3045m)
III+
II

Perched far above the plains of Turin, its iconic position and relatively easy summer access ensure Rocciamelone remains one of the most popular high summits in the Alps. Every year on 5 August, crowds of pilgrims make the ascent for a special mass held on the summit – the festival of Notre Dame des Neiges.

The first known ascent of Rocciamelone was made by Bonifacio Rotario d'Asti in 1358. That same year, he commissioned a small gilt bronze triptych of the Virgin to be housed in a small chapel built at the summit. This artwork, executed by a Flemish workshop, was finally brought down to the cathedral of Susa in 1673 and is now displayed in the museum.

In 1899, the sculptor Giovanni Antonio Stuardi modelled the giant bronze statue of the Virgin that adorns the summit. Cast in Milan, the four-meter-high statue weighs 650kg. It was broken down into eight pieces to transport to Susa, where it was temporarily reassembled and blessed. At that time, there was no path up the final steep bastion, so 60 soldiers from the local Alpine regiment were tasked with manoeuvring the cumbersome pieces to the summit, along with 800kg of materials for the support structure.

Above: Valentine thinking about chocolate. Top left: inside Rifugio Santa Maria. Bottom left: Rifugio Santa Maria at the summit of Rocciamelone.

24 Bivacco Beltrando (2439m)

Alpe Losa (1230m), Frassinetto, Pont Canavese, Italy

7 places (8 mattresses, 12 blankets)

Woodstove, but no hobs or cooking equipment

Spring by approach path at c.1600m

3h30. +1200m & 5km from Alpe Losa (Hiking – T4)

From a small parking area just below Alpe Losa (Frassinetto), follow path 8 leading E to I Piani. Cross Rio Cialma to reach the last water source, a piped spring at c.1600m. Continue past the ruins of Alpi Livi, ascending the WSW ridge of Punta Rama. The final 300m to the summit (2439m) and bivouac have sections equipped with cables.

Punta Rama (2439m) Ski Tour – 7km, +1600m (Ski – 2.3)

While the SW approach from Frassinetto is complicated by steep terrain in winter, it is possible to reach Punta Rama on skis from Val Chiusella to the north. From Fondo (1076m), ascend the road SE towards the disused lifts of Vaudanza (1250m). Climb S into the Rio Piera valley and follow this to the col between Cima La Rubbia (2427m) and Gran Munt (2235m). From the col, ski SW then ascend WSW to reach Passo Savenco, between Cima di Pal (2494m) and Punta Rama. A short scramble S along the crest leads to Punta Rama (5hrs from Fondo).

Punta di Verzel (2405m) via Cresta Beltrando (Hiking – T4)

From Bivacco Beltrando on Punta Rama, follow the cable-equipped crest SE over Monte Cavallo to Colle d'Asni. From the col, it is possible to escape by descending rough hillside SSW to join a path at Beccua. To continue along the ridge, remain on the crest over Punta Prafourà to reach Punta di Verzel (2hrs from Bivacco Beltrando). From the summit, descend more easily via the SW ridge to a shoulder at 2246m. From there, either return directly to Alpe Losa via a path that leads NW past Chiona, or continue the ridgeline circuit SW over Punta Quinseina (2312m), descending its W ridge to Pian Col (1346m). From there, follow a path N to rejoin the road leading back up the valley to Alpe Losa.

Perched on the summit of Punta Rama at 2439m, Bivacco Beltrando is a striking example of Italy's new wave of high mountain architecture. Set on the watershed between Val Soana and Valle Sacra, on the eastern edge of the Graian Alps, this unmanned shelter combines bold design with robust functionality. Its panoramic position offers sweeping views from Monviso to the Gran Paradiso massif, Monte Rosa and the plains of northern Italy.

Commissioned by the municipality of Frassinetto and funded by the Regione Piemonte, the bivouac, together with the equipped ridge leading to Punta di Verzel, is dedicated to Marco Beltrando, a respected alpine geologist, researcher at the University of Turin, and passionate mountaineer from Ivrea, who lost his life in these mountains in 2015. The project was conceived to promote tourism in the area.

Top: David Lee on the approach to Punta Rama.
Right and opposite: exterior and interior views of Bivacco Beltrando.

The construction was demanding: 40 days of work were split between prefabrication in the local sawmill and a delicate high-altitude assembly on the summit of Punta Rama. The structure is built to withstand the extreme weather typical of such elevations, using sustainable materials wherever possible. Locally sourced poplar, larch and ash were treated with nanotechnology to enhance durability, while the aluminium exterior offers long-term resistance to wind and water.

Despite its remote and rugged location, the bivouac is remarkably well equipped. The wedge-shaped structure features an entrance space with three large east-facing windows designed to capture spectacular alpine sunrises. Behind this lies the compact dining area, with two lower bunks and a mezzanine offering bed space for five. Visitors will find mattresses, blankets, electric lights and USB charging ports. For cooking, hikers should bring their own gas stove.

Although exceptionally well insulated, the bivouac intriguingly includes a wood-burning stove, despite the nearest trees growing some 800m below. When we visited, a modest supply of wood had been stocked in the shed. Winter access is rather more demanding, and anyone attempting the ascent in colder months should be prepared to carry their own fuel.

Though places cannot currently be reserved, hikers are encouraged to coordinate visits via the bivouac's Facebook page to reduce the risk of overcrowding. Reaching the bivouac requires a strenuous hike, with the most direct approach starting from Alpe Losa, a small trailhead accessed by a partly unpaved road from Frassinetto in a side valley of Val Soana. The final 300 vertical metres of the ascent are equipped with fixed ropes and metal steps, so a via ferrata kit and helmet are advisable.

Above: sunset over Gran Paradiso, seen from Punta Rama. Previous spread: David on the Cresta Beltrando.
Top left: David leaving Bivacco Beltrando. Bottom left: inside the hut.

25 Bivouac d'Estelette – Adolfo Hess (2958m)

Frêney (1583m), Val Veny, Italy
or Chapieux (1550m), La Vallée des Glaciers, France

4 places (No mattresses, 6 blankets)

No fireplace or permanent stove

Melt snow or collect water on approach

6h00. +1400m & 10km from Frêney (Hiking – T3)

From Frêney, follow the track to Rifugio Soldini, then head W across the moraine and ascend the couloir leading to Col d'Estelette (snow or rubble, yellow paint marks). From the col, briefly head W up the crest before descending 50m N to reach the bivouac. Alternatively, from Chapieux (France), take the bus to Chalets des Lanchettes and ascend the path to Col de la Seigne. Traverse NW to the remains of Glacier de l'Estelette. Cross the moraine, and join the previous route.

Aiguille des Glaciers – Arête Kuffner / SE Ridge (900m, D-, IV+)

From Bivouac d'Estelette, head W along the ridge, passing L of a tower and then follow the crest down to the Glacier de la Lée Blanche. Head W across the glacier to the rocky bastion at the base of the main ridge. Bear L and ascend slabs and a groove on the L (SW) flank to reach a notch on a spur that descends SW. Climb the spur for several pitches (III) to gain the main crest (snow in early season), leading to the base of a steep red tower. Bypass this via a chimney on the L (IV, or easier via slabs further L), then head back R behind the tower. Follow the easy crest of rock or snow, with occasional steeper pitches and variable rock quality, to reach the crux tower below the summit. Either i) climb the crack directly above the terrace (V+), or ii) traverse R (N) via a horizontal step, climb a chimney for 5m, and regain the top of the V+ crack by a large step back L (IV+). On the next 30m pitch, ascend via a small overhang (IV+). A final 30m pitch (IV) leads to the top of the tower. Continue upwards to the summit (7hrs from Bivouac d'Estelette). Descend via the N ridge and W flank.

Bivouac d'Estelette – Adolfo Hess is located by the Col d'Estelette, near the base of the southeast ridge of the Aiguille des Glaciers (3816m). The hut primarily serves as a base for this climb, known as the Arête Kuffner, first ascended in 1887 by the Austrian industrialist Moritz von Kuffner with Swiss guides, Alexander Burgener and Josef Furrer. Positioned on the Franco-Italian border, the Aiguille des Glaciers is the most southerly major peak in the Mont Blanc massif.

Built in 1925 by the Italian Academic Alpine Club (CAAI), Bivouac d'Estelette was the first of the "Ravelli"-type bivouacs to be installed in the Alps. Among its earliest visitors were Italo Brosio and Luigi Bergera, who claimed the "first complete ascent of the southeast arête."

Above: Tom Coney and Valentine cycling up Val Veny.
Right: Tom arriving at Bivouac d'Estelette. Opposite top: departing from the hut at dawn.
Opposite bottom: approaching the final tower on the SE ridge of the Aiguille des Glaciers.

In 1925, they climbed the steep terminal buttress of the ridge that Kuffner had bypassed via a pendulum to the right. Brosio wrote: "The fixed bivouac of the CAAI on the Col de l'Estellette is truly providential: it is a safe, perfectly comfortable shelter; a peaceful nest on the edge of an awesome precipice. Situated on the rocky cliff that separates the Ghiacciaio dell'Estellette from the tormented torrent of ice of the Allée Blanche, its beautiful position in the heart of the Trélatête group rewards the effort of the ascent. The close-up view of the west side of Mont Blanc is marvellous, and the eye can also sweep far over the superb circle of the Western Alps, especially after ascending the easy half-hour climb up the nearby Aiguille d'Estellette."

Since 1951, the hut has been dedicated to Adolfo Hess, a CAAI founding member. Along with Mario Borelli and Francesco Ravelli, Hess designed and installed many of these early high-altitude bivouacs, adapting the concept of sheet metal shelters used during the First World War.

The structure is a low, half-cylinder made of wood with galvanised steel cladding. Standing just 125cm high, it resembles a pigsty and accommodates only four people. Like the Bivacco Craveri on the Peuterey ridge, it appears austere by modern standards. Thirty metres above, there are wooden remains of an earlier hut.

Above: Tom and Valentine approaching the final tower of the SE ridge of the Aiguille des Glaciers.
Right: Aiguille des Glaciers at sunrise from Testa del Rutor. The red line shows the SE ridge and Bivouac d'Estelette (circled).

Aiguille des Glaciers (3816m)
IV+
V+
III
III
IV
III

Under a burning July sun, Tom, Valentine and I cycled up Val Veny, bouncing along the rocky track against a tide of Tour du Mont Blanc hikers sweating under laden packs. By late afternoon, we reached Col d'Estelette, where Bivouac d'Estelette stood out vividly, freshly repainted bright green and gleaming like a giant beetle.

We cooked inside, hunched under the low roof, thankful that no other teams had arrived to squeeze into the cramped space. Gazing out through the tiny door at sunset, we watched shifting clouds catch the last rays as the summit of Mont Blanc briefly blazed vermilion. As dusk deepened into violet hues, we settled into our sleeping bags.

The following day we set out at five, pausing for slow-shutter photos in the gloaming, then raced along the lower ridge to capture the crisp blue tones of dawn. Layers of clouds spread up from the valley to embrace the sky. The sun crept above the Peuterey ridge for a couple of minutes before being harassed by the cumulus. Wandering up the glacier with a heavy camera on my hip, I felt cheated by the brief promise of a stunning sunrise.

Starting hesitantly up the main ridge, I wondered if the day was turning against us. After several interesting pitches, we emerged onto the upper crest and trotted to the foot of a steep, red tower. Above, the cloud base descended and enveloped us, and the ridge beyond began to feel endless. Hours passed scrambling through chilling fog, climbing rocky steps and traversing snow blades until finally, a steep buttress emerged from the mist.

The rock was damp. As Tom and I debated a way around the imposing cracked wall, Valentine tackled it directly. In minutes, she forced her way up and called for us to follow. A couple of sturdy pitches later, the ridge levelled and began descending. Still shrouded in mist, with only 20m of visibility and a biting wind, we started down the crenellated north ridge, without so much as a pause. It was with no little relief that we reached the snow, emerged below the cloud layer and began galloping and glissading down the glacier back to the valley.

Above: Valentine and Tom descending the Aiguille des Glaciers via the N ridge.
Top left: Tom and Valentine inside Bivouac d'Estelette. Bottom left: Valentine and Tom on the lower SE ridge at sunrise.

Ben Tibbetts is an adventure photographer, filmmaker, artist and internationally certified mountain guide (IFMGA).

Ben studied art at the University of Edinburgh and the Geneva School of Art and Design. His images have won numerous awards, and his films have been selected in festivals all over the world.

With a passion for wild and remote environments, he has climbed and skied new routes in Greenland, Antarctica, Kyrgyzstan and the Alps and worked as a guide in the most isolated corners of Antarctica. In 2018 Ben became the second British person to climb all 82 of the 4000m peaks.

Valentine Fabre Valentine Fabre is an adventurer, climber and ski mountaineer. She works as an emergency and sports doctor. A former member of the French national ski mountaineering team, she is a two-time World Champion in the discipline.

Now an aspirant IFMGA mountain guide, she has climbed hundreds of technical routes in the Alps and been on numerous expeditions to the greater ranges.

In 2021, Valentine and Hillary Gerardi set the fastest women's non-stop ski traverse of the Haute Route, covering the 107km and 8000m of ascent from Chamonix to Zermatt in just 26 hours.

ACKNOWLEDGEMENTS

We would like to thank all the people with whom we have shared these adventures and the guides, local experts and editors who generously supported the making of this book.

Edited by: Valentine Fabre, Julie Horsfall, Benjamin Hall, Tony Moody, Martin Block, Marcea Ennamorato, Moira Hunter, Simon Fitall, Finlay Wild, Sarah d'Adamo, Nick Boyd, Mike Arpaia, Brian Clarke, Sara Tibbetts, Misha Gopaul, William Mitchell, Alex Candida, Saro Costa, Cain Olsen, Stuart Johnston, Michael Devor, Lara Kesterton, Manon Stravens, Peter Gerard, Jon Morgan, Andy Helm, Eric Vola & Victor Saunders.

With thanks to: Matteo Canova, Stefano Girodo, Luca Gibello, Roberto Dini, Bruno Bonino, Michel Coranotte, Paul Roumegoux, Valentina Eula, Andrea Cassi, Luca Signorelli, Piero Bertotto, Devis Guiguet, Benoît Dugardin, André Imbert, Yves Saunier, Isabelle Curt, Christine Baccon, Marc Bourgeat, Eric Pulvin, Marie-Thérèse Donzel-Gargand, Paul Mathevon, Clément Ohresser, Nicolas Genans-Boiteux, Jean-Bernard Challamel, Alex Strohl and the dedicated staff of the Alpine Club, CAI, CAS, and CAF.

I would also like to thank my equipment partners for their continued support: Rab, Petzl, Scarpa, Ski Trab and Julbo.

First published by Alpenglow Editions in 2025.

ISBN: 978-1-9161231-3-7

Photography and text by Ben Tibbetts unless otherwise stated.

For more information please visit: **www.bentibbetts.com**

MIX
Paper | Supporting responsible forestry
FSC® C106600

Ben Tibbetts has asserted his rights under the Copyright, Designs and Patents Act 1988 to be identified as author of this work.

Map data courtesy of OpenStreetMap.

Printed and bound in Slovenia by Latitude Press Ltd.

A CIP catalogue record for this book is available from the British Library.

Disclaimer: Mountaineering is an inherently dangerous activity. Participants should understand the risks involved and be responsible for their actions and safety in this environment. Although every effort has been made to obtain current information before publication, omissions are inevitable. If you discover any errors or changes to an itinerary please contact us by email – ben@bentibbetts.com.

The authors cannot be held responsible for accidents that may occur on any of the routes described, or in relation to any information presented in this book.

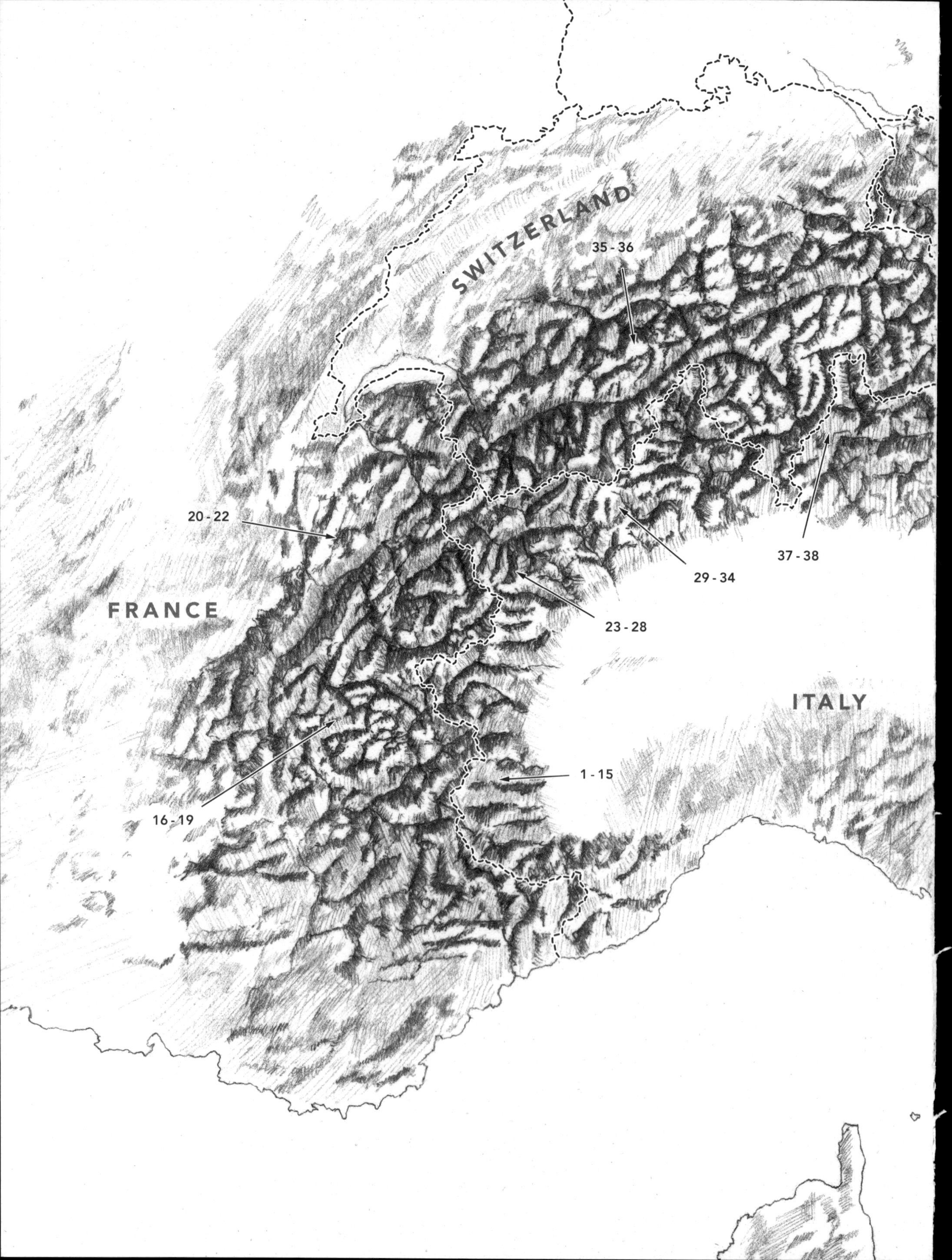

SWITZERLAND
35 - 36
20 - 22
FRANCE
29 - 34
37 - 38
23 - 28
ITALY
1 - 15
16 - 19